ELECTROTECHNOLOGY
VOLUME 4

HEAT PUMP TECHNOLOGY

ELECTROTECHNOLOGY VOLUME 4

HEAT PUMP TECHNOLOGY

NORMAN W. LORD
ROBERT P. OUELLETTE
METREK Division of the MITRE Corporation
McLean, Virginia

PAUL N. CHEREMISINOFF
New Jersey Institute of Technology
Newark, New Jersey

ANN ARBOR SCIENCE
PUBLISHERS INC
P.O. BOX 1425 • ANN ARBOR, MICH. 48106

230 Collingwood, P.O. Box 1425, Ann Arbor, Michigan 48106

Library of Congress Catalog Card No. 79-56112
ISBN 0-250-40341-2

Manufactured in the United States of America

PREFACE

This book continues the series of studies initiated and sponsored by Electricité de France surveying the status of electric technologies. The objective of this book is to promote technology using electricity and to diffuse information favorable to its development.

By conveying heat from an adequate thermal reservoir at a rate several times that of the mechanical energy used by the engine, the heat pump can compensate for most of the fuel combustion energy lost in the electricity generation process or, if mechanical energy is supplied directly by an independent prime mover, realize a heating energy bonus through use of engine exhaust heat. Heat pump technology in the United States is based on refrigeration cycles and is normally used to produce a system that can either heat or refrigerate an enclosed region by extracting heat from, or exhausting heat to, an exterior region acting as a thermal reservoir or sink.

Early heat pumps were little more than air conditioners with a reversing valve to interchange thermodynamic roles of the indoor and outdoor coils. Such units proved unsatisfactory because of the following departures from simple air conditioning requirements: (1) greater required operating temperature range, particularly at low temperatures in the exterior coil; and (2) longer continuous operation periods. These problems were overcome with a more robust design and special control measures that ensured continuous refrigerant flow through the thermodynamic cycle. The new heat pumps can be designed to operate with either air or water as ambient medium at exterior and interior heat exchangers. However, as it is somewhat more costly to exchange heat between refrigerant and water, most heat pumps (particularly in the residential market) are designed to operate with air as the ambient at both exchangers.

Current approaches to improve heat pump effectiveness are aimed at:

- increasing efficiency of refrigeration loop through reduction of heat leaks and fluid pressure loss;
- improving thermodynamic quality of thermal reservoirs; and
- improving capacity modulation to conform better with HVAC requirements.

Improvements in the refrigeration loop efficiency are organized primarily by work on the separate components of the system, chiefly the compressor, which is being improved steadily to reduce the following losses:

- valve, valve port, suction and discharge tube, and muffler pressure drops;
- internal leakage of discharge gas to suction side of compressor;
- friction, hermetic motor and compression cylinder conduction heat losses to suction gas;
- valve inefficiencies (mechanical);
- oil circulation work loss;
- clearance volume (compressed gas reexpansion loss); and
- anisentropic compression.

System design improvements are being approached primarily through improving the management of external reservoirs. Relatively small thermal reservoirs, based on either a water or ceramic storage medium, are being used to modulate the duty cycle of the heat pump. These are often assisted by solar collection devices which, in this application, can work effectively at moderate temperatures.

Heat pump use is being promoted by sponsored programs of the U.S. government and utilities. The following general benefits are anticipated:

- for home heating, a shift from gas and oil to coal-generated electricity;
- more effective use of gas in gas-fired heat pumps; and
- better balancing of utility loads.

Distinct from gas utilities, electric utilities are interested in heat pumps for load-balancing and as a way to dissuade electric heat consumers from switching to gas or oil as electricity rates rise. Gas supply utilities have only recently begun to sponsor gas-fired heat pump developments along with the U.S. Department of Energy (DOE) as a move to counter any shift to electric heat pumps on the part of their own customers. On the other hand, electric utilities are managing an impressive array of programs, all carefully tailored to the specific benefits that can be gained for the utility and its customers in their particular service area.

U.S. government-sponsored research in heat pumps is carried on primarily by DOE, with some effort by the U.S. Environmental Protection Agency (EPA) and National Aeronautics and Space Administration (NASA). DOE is involved in joint sponsorship of both gas and electric utility programs, and their own efforts, which are meant to implement the utility efforts in the following ways. Whereas utilities are interested in new technology that will encourage heat pump use, DOE's work centers either on exhaustive state-of-the-art assessment on purely off-the-shelf units, or on attempts to develop major advances in the technology with concerns over economic uncertainties temporarily deferred. This is in contrast to the utilities' experiments with diurnal thermal storage, which can be fairly cheap and simple, as exemplified by DOE's Oak Ridge National Laboratory advanced annual thermal storage concept.

This book attempts to cover these aspects and provides further references and contacts.

Norman W. Lord
Robert P. Ouellette
Paul N. Cheremisinoff

Norman W. Lord is a broad-spectrum physicist with the MITRE Corporation whose research areas include the structure of solids, ocean acoustics and atmosphere-land interactions. He has consulted extensively for government and industries on a wide variety of problems in operations and estimating technological trends.

He received a BEE from Brooklyn Polytechnic Institute (now New York Polytechnic), then, after military service, earned his MA and PhD in Physics from Columbia University. Dr. Lord has been a research physicist at Johns Hopkins University Applied Physics Laboratory, Hudson Laboratories of Columbia University and Travelers Research Center in Hartford. His consulting work has included a variety of specific industrial technology problems which span practically every major manufacturing industry group. He has written more than 40 technical papers covering these areas and his research.

Robert P. Ouellette is Technical Director of the Environment Division of The MITRE Corporation. Dr. Ouellette has been associated with MITRE in varying capacities since 1969 and has been Associate Technical Director since 1974. Earlier, he was with TRW Systems, Hazelton Labs, Inc. and Massachusetts General Hospital. He graduated from the University of Montreal and received his PhD from the University of Ottawa. A member of the American Statistical Association, Biometrics Society, Atomic Industrial Forum and the NSF Technical Advisory Panel on Hazardous Substances, Dr. Ouellette has published numerous technical papers and books on energy and the environment. He edited the comprehensive survey, *Electrotechnology*, published in 1978 by Ann Arbor Science Publishers, Inc.

Paul N. Cheremisinoff is Associate Professor of Environmental Engineering at the New Jersey Institute of Technology. He is a consulting engineer and has been a consultant on environmental/energy/resources projects for MITRE Corp. A recognized authority on pollution control, he is author/editor of many publications, including several Ann Arbor Science handbooks on pollution and energy, such as *Pollution Engineering Practice Handbook; Carbon Adsorption; Environmental Impact Data Book; Industrial & Hazardous Wastes Impoundment; Environmental Assessment & Impact Statement Handbook*. Mr. Cheremisinoff is also Engineering Editor of *Water & Sewage Works*, an international journal.

TABLE OF CONTENTS

LIST OF FIGURES

LIST OF TABLES

CHAPTER 1

HEAT PUMP SUITABILITY FOR RESIDENTIAL HEATING

INTRODUCTION

The heat pump, used in the United States, transfers thermal energy both into and out of a closed space that is required to have a temperature stabilized against weather variations. Because winter-summer extremes of outside temperature require both heating and cooling, the normal economic alternative is to have single systems serve to either cool or heat, with auxiliary switching mechanisms effecting the reversal.

The central thermodynamic principles used in such systems are as follows:

- Adiabatic compression raises the temperature of a heat transfer fluid so that it expels its internal heat through a first exchanger.
- Adiabatic expansion of the fluid occurs, usually assisted by a liquid-gas phase change, so that it cools and thereby extracts heat from a second exchanger.

The fluid is then conveyed to a compressor for a repetition of the cycle. When the first exchanger (at output of compressor) is inside the space to be temperature-stabilized, the system operates to heat it against a cool ambient exterior. The second exchanger is then usually outside or in a sheltered, but uninhabited, space, and heat is being extracted from the exterior as a thermal reservoir. The sheltered space must exchange air readily with the exterior, but not with the closed inhabited space. In the opposite mode, which can be achieved by reversing circulation of the transfer fluid, heat is extracted from the interior exchanger and expelled through the exterior exchanger. In both modes, work is done to extract heat from a low-temperature reservoir and supply it to a high-temperature reservoir.

The possibility that the heat energy output thereby transferred could greatly exceed the mechanical or electrical energy input has provided continuing interest in the device. Although the underlying thermodynamics were developed in the mid-19th century, applications were slow to develop. Some equipment suited to the provision of industrial process heat was manufactured in Switzerland in the 1920s. However, in the U.S., the earliest serious attempt to apply a heat pump for space heating occurred in 1934. The offices of the Atlantic City Electric Company, Salem, New Jersey, were heated by a plant that consisted of four 5-hp compressors. An overall coefficient of performance (COP) of 3.3 (instantaneous cycle COP was 3.9) was obtained. In the years immediately following, a number of comparable installations were made of sizes up to 50 hp. Through 1947, it is estimated that the total number of heat pump installations in the United States was between 300 and 450, the size range of these devices running from 1.5–300 hp.

ELECTRIC HEAT PUMP SYSTEMS

Electric heat pump systems, using an electric motor to drive the compressor, air fan, water pumps and other ancillaries, are the predominant types used in the U.S. Key components of such systems are shown in Figure 1-1. The main function of the compressor is to pump a refrigerant vapor from a relatively low suction pressure to a higher head pressure. The suction and head pressures that a compressor experiences are a function of the design, ambient temperatures and possible abnormal fault situations that develop during the life of a heat pump system. Under

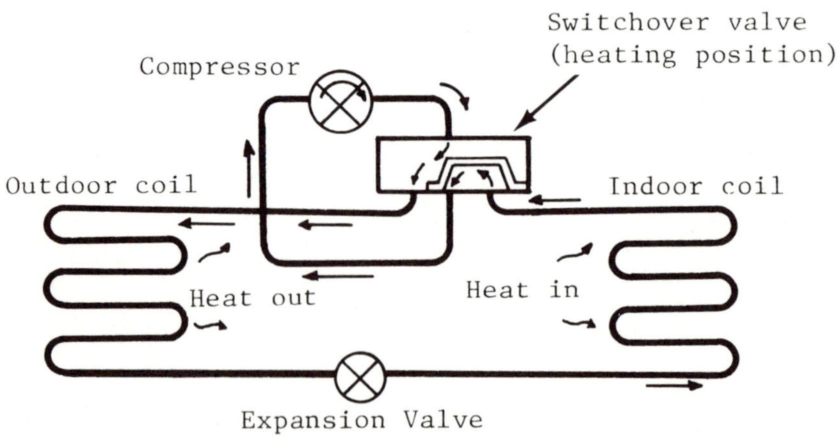

Figure 1-1. Simplified heat pump flow diagram in cooling mode showing use of switchover valve to reverse flow of refrigerant (Pickelsimer, 1977).

cooling conditions, the refrigerant is pumped from the compressor through the condenser or outdoor coil. On leaving the outdoor coil in liquid form, the refrigerant travels through the evaporator or indoor coil. Here, as it changes from a liquid to a gas, it removes heat from the interior, thereby cooling the indoor coil. The gas is then drawn back into the compressor, where it is compressed again into a very high-pressure hot gas that is liquefied or condensed in the outdoor coil.

The first heat pumps were more or less simplified versions of this cooling unit and were obtained by merely reversing the flow of refrigerant, as shown in Figure 1-1. With the aid of a switchover valve, the outdoor condensing unit actually became the evaporator, and the indoor evaporator became the condensing unit. Therefore, the hot gas was pumped indoors, where it gave off heat as it was condensed into a liquid in the indoor section. The liquid was then pumped into the outdoor section, now the evaporator, where the refrigerant changed back into a gas, absorbing heat from outdoor air during this process.

After World War II there was a strong promotional effort by electric utilities to capture the residential heating market. Fuel was cheap, and electricity represented "better living" in keeping with higher standards of personal comfort. Heat pumps could alleviate the extra cost, and air conditioning manufacturers marketed their units or heat pumps with little more than the reversing switch added. This concept led to widespread operational failures that severely hampered the introduction of heat pumps until the mid-1960s. The failures were due principally to the following:

1. A heat pump must operate over a much wider range of conditions than are met in air conditioner operation.
2. For a heat pump, the thermal exchanger at the suction end may be at temperatures as low as 10°F (-12°C), while an air conditioner will typically have its cooling unit no lower than 50°F (10°C).
3. The necessary "run" time for a unit in heat pump service may easily be five times as long as the "run" time needed in air conditioning service.

These difficulties produced multiple failures that were related in various ways to low-temperature malfunctions. Frequently they had disastrous consequences. Additionally, the wider operating range required seems to cause some extra compressor wear and tear under mild heating conditions: the machine may not run too long, but it has to do its job at slightly too high a pressure. Despite the problems–general deterioration due to wear and tear and failures induced by the severe operating conditions–heat pump usage grew slowly during this early period.

In 1957, residential heat pumps installed totaled about 26,000. By 1967, the total had risen to 400,000, reflecting an installation rate of 20,000-30,000/yr at the beginning of the period and about 50,000/yr in the latter 1960s.

In present-day heat pumps, the specific defects of the earlier generation have been eliminated. There are four basic system designs centered on the constituency of the thermal reservoir used as heat source and sink exterior to the temperature-stabilized interior space, and also intermediate thermal transfer systems used inside:

1. The first is air as the heat source-sink and air as the direct interior heating and cooling medium. By far the largest proportion of presently available heat pumps fall into this category.
2. The second is air as the heat source-sink and water as the intermediate interior heating and cooling medium. Few units are made in this category. Heat exchange to air at interior temperature (60-80°F) is too simple and cheap for any other arrangement to compete against economically.
3. The third is water as heat source-sink and air as the heating and cooling medium. While still a minority compared to air-to-air, this category is increasing in importance in areas where a water thermal reservoir is available.
4. Water as heat source-sink and water as the heating and cooling medium is the fourth. This category is not yet as popular as water-to-air.

Each of these basic designs can supply the required heating and cooling effect by changing the direction of the refrigerant flow or by maintaining a fixed refrigerant circuit and changing the direction of the heat source-sink media. A third alternative is to incorporate an intermediate transfer fluid into the design. In this case, the direction of the fluid is changed to obtain heating or cooling, and both the refrigerant and heat source-sink circuits are fixed. The fixed refrigerant circuit designs, generally referred to as the indirect type of application, are becoming increasingly popular, particularly in the larger capacities.

Heat pump cycles lend themselves readily to a variety of conservation measures, and even if total heating by one heat pump is impractical, it can be used to provide household hot water during most of the year and to supplement the available heat from a solar or thermal storage system. A large enough hot water heater can also serve as a thermal reservoir in systems that program heat pump operation for utility load balancing or higher seasonal efficiency. A schematic of an arrangement accommodating such possibilities is shown in Figure 1-2. Of course, in a practical arrangement there may be elaborate auxiliary control to bypass various parts of the system and maintain its operation in the most efficient accessible state.

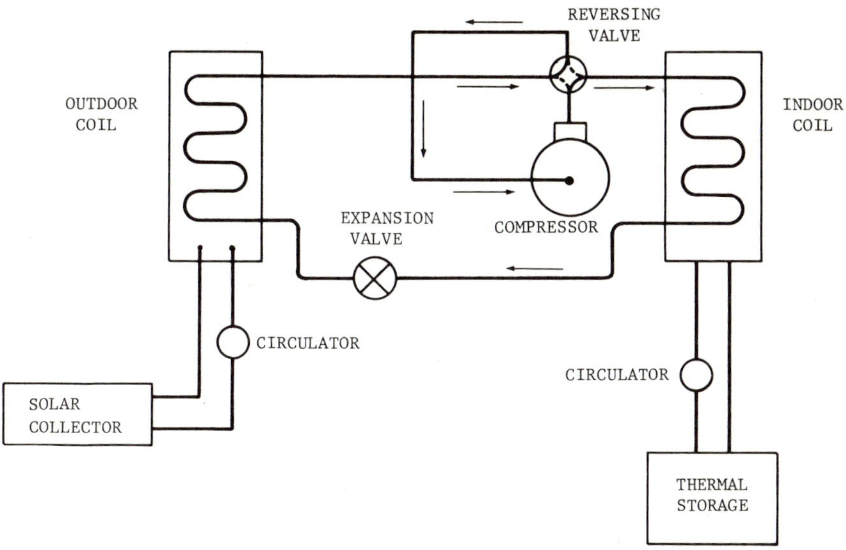

Figure 1-2. Prototype heat pump arrangement.

The heat pump coefficient of performance cannot be related usefully to the efficiency of heating in the conventional fossil fuel sense. Such an efficiency is defined as the desired effect divided by the energy required to produce it. For a fossil fuel furnace (Figure 1-3), the efficiency is:

$$\text{Eff} = \frac{W}{Q_H} \times 100\%$$

where W is the heat supplied to the interior, burning fuel to release Q_H, but also exhausting Q_L through the flue. In practice, Q_L can be kept to less than one-third Q_H. If we wanted to use electric resistance dissipation as an alternate heat source, we would be faced with the fact that electricity is generated with one-third efficiency, at best, so that fossil fuel burning is at least twice as cost-effective (assuming equal fuel costs at home and power plant as an energy basis).

The electric heat pump can overcome this inherent disadvantage by using electrical energy to transfer thermal energy from a free reservoir. We then have no useful way to rate the system by efficiency, but rather in terms of the ratio between the energy transferred and the electrical work required to do this.

In the heating cycle, the COP_h of the heat pump is, by definition, the desired effect (heat into the home, Q_H) divided by the energy required to produce it (electrical input = W) as shown in Figure 1-4, or:

$$COP_h = \frac{Q_h}{W}$$

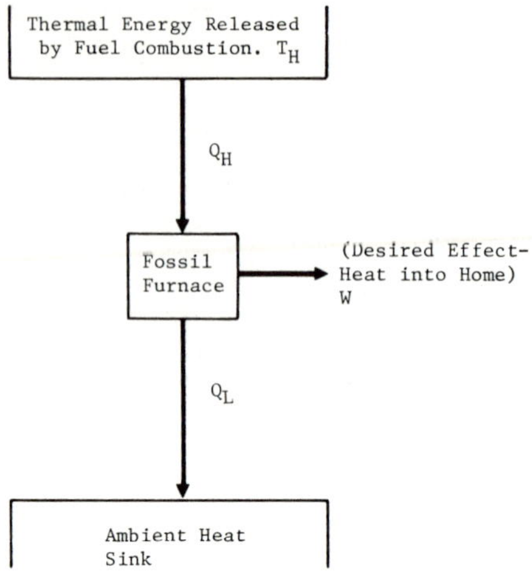

$$\text{Efficiency} = \frac{W}{Q_H} \times 100\% \quad 100$$

Since $W = Q_H - Q_L$,

We seek to minimize Q_L in order to maximize W.

Figure 1-3. Efficiency (Pickelsimer, 1977).

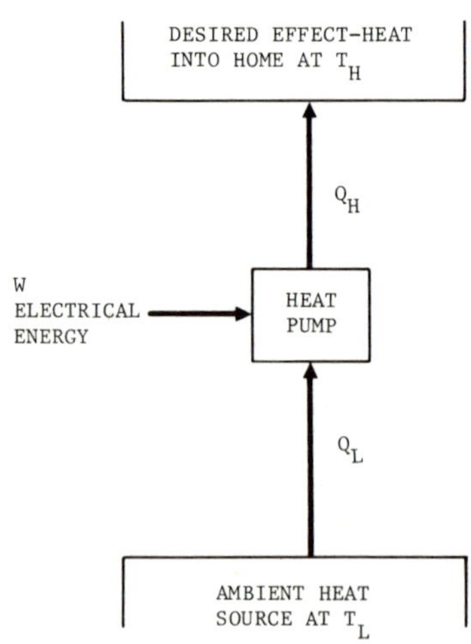

$$COP_h = \frac{Q_H}{W}$$

Figure 1-4. Coefficient of performance (COP_h), heating mode.

For the heat pump, we seek to minimize $W = Q_H - Q_L$ by maximizing Q_L relative to Q_H. Ancillary drive power will also contribute to W, and we will be concerned with the capacity represented by the maximum Q_H, which can be achieved mainly through the compression evaporation cycle. In any case, it is Q_H that is of primary interest and accessible to measurement.

For the cooling cycle, the heat pump COP_c is defined as the desired effect (heat removal from the home, Q_L) divided by the energy required to produce it (electrical input = W) as shown in Figure 1-5, or:

$$COP_c = \frac{Q_L}{W}$$

Again, we want to maximize Q_L relative to Q_H; however, this time it is Q_L (interior cooling) that is of primary interest and accessible to measurement.

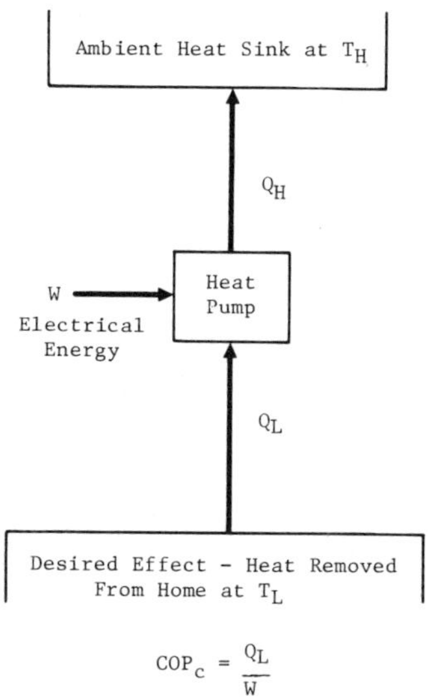

Figure 1-5. Coefficient of performance (COP_c), cooling mode (Pickelsimer, 1977).

COMPARISON WITH GAS AND OIL HEATING

In the case of COP_H, heating coefficient of performance, we can see that if this is two or greater, the inherent disadvantage of electricity for heating is relieved. We are then left primarily with an economic assessment hinging on local factors to determine whether the electric heat pump is a financially sound investment. These factors will be local costs of fuel vs electricity and relative capital investment required (air conditioner plus furnace vs heat pumps and maintenance costs). Further, there are the problems of operational capacity limits. The combustion of fuel for heating maintains the same economy no matter how low the temperature within normal atmospheric limits. Mechanical work to extract heat from a chilly atmospheric environment, however, becomes increasingly uneconomical and eventually unfeasible, as temperatures drop well below freezing. In this latter range, the heat exchanger for an air-source heat pump will frequently condense an excess of ice on the vanes so that air passage across the coil is occluded. There is then sharply reduced heat transfer from the outside air, the refrigerant gas may freeze under suction, and it becomes imperative to defrost the vanes by a costly (albeit temporary) reversal of the heat pump's operation.

If a furnace must be retained for cold spells or even routinely subfreezing nights, then its capital cost adds considerably to the economic burden of the heat pump system. As a relatively new technology, heat pump costs and maintenance are necessarily higher than those of an air conditioner. Hence, its COP_H has to be high enough to give it an operational advantage (electricity vs fuel) great enough to repay the investment before the system must be totally replaced.

REGIONAL VARIATION

Outside temperatures will vary, so auxiliary heaters must be used under some circumstances. As defrosting is required for air-source heat pumps, the performance obtained from a real system will, in general, be less than the COP (instantaneous COP) under nominal rating conditions.

Taking the cycle COP as heat recovered divided by work input to the compressor, or better, the COP for the apparatus (overall COP) as the net heat effect divided by the total power consumed correctly indicates the heat gain available from a properly designed work cycle. If appropriate standard conditions are selected, different heat pump systems may be compared.

The "overall" COP includes power consumed by auxiliaries, such as fans and circulating pumps. Auxiliary requirements may amount to about 10% of the compressor power. However, except in a case such as the recovery of process heat to preheat an incoming stream, where both source and sink are apt to remain at, or near, some definite temperature level, COPs under a single set of rating conditions are not sufficient to determine the time average value of the system's performance. The Seasonal Performance Factor (SPF), a more direct measure, is taken to be the energy actually used over an entire season, including the heat pump, all auxiliaries and any supplemental heat that may be needed, divided into the energy that would have been used (expressed in consistent units) if no heat pump were available.

The SPF covers every conceivable deficiency in an installed heat pump system. The available work/heat cycles fall short of what is ideally obtainable and, in addition, we must accept certain mechanical losses in the operation of any real device. As well as the inevitable losses that must be associated with steady operation, we have to allow for the fact that a working pump (air-to-air) may occasionally have to be run in a defrost mode to deice the evaporator coils—energy that is not used for heating. Finally, an auxiliary heat source, generally electric, must be used under some conditions to supplement the heat that can be delivered by the pump.

In general, well-designed machines (documented in Chapter 4) currently have overall COPs of 2.4-3.1 under mild standard operating conditions [45°F (7°C) outside, 70°F (21°C) inside]. If such conditions were met throughout the heating season, the SPF would be well above 2. However, through most of the winter conditions can be much worse for part of every day, depending on the regional climate. Moreover, because the heating requirement during such periods is much higher, the low instantaneous COP for such conditions will depress the seasonal performance factor considerably. Typically, as temperature falls below 17°F (-8.33°C), the HVAC industry's low ambient temperature standard, instantaneous COP falls below 1.5, and the heat pump provides little better than resistance heating.

Therefore, we must anticipate that the value of using a heat pump compared to fuel burning, or even electric resistance heating, will vary widely over different regions of the United States. To gain some idea of the characteristics of this variation, we must consider the regional climatic variations, particularly with respect to diurnal temperature extremes. Figure 1-6 indicates the states with time zones. To show winter operating conditions, Figures 1-7 and 1-8 display maximum and minimum daily temperatures for January. From these charts we have delineated five regions within which heat pump winter operating conditions will be relatively uniform.

1. North Mountain and North Central. With daily maximum temperatures in the 20s and daily minimum temperatures around 0°F, heat pumps would be almost useless in January.
2. A wide band stretching from South Mountain (New Mexico) to Northeast and Southern New England. With daily maximum around 40 and minimum around 20, heat pumps here are of marginal benefit in January and may not be economical against readily available alternatives.
3. Deep South, Texas to Georgia and South Carolina, including Florida. Where daily maximums are 65°F and minimums, 45°F, when heat is

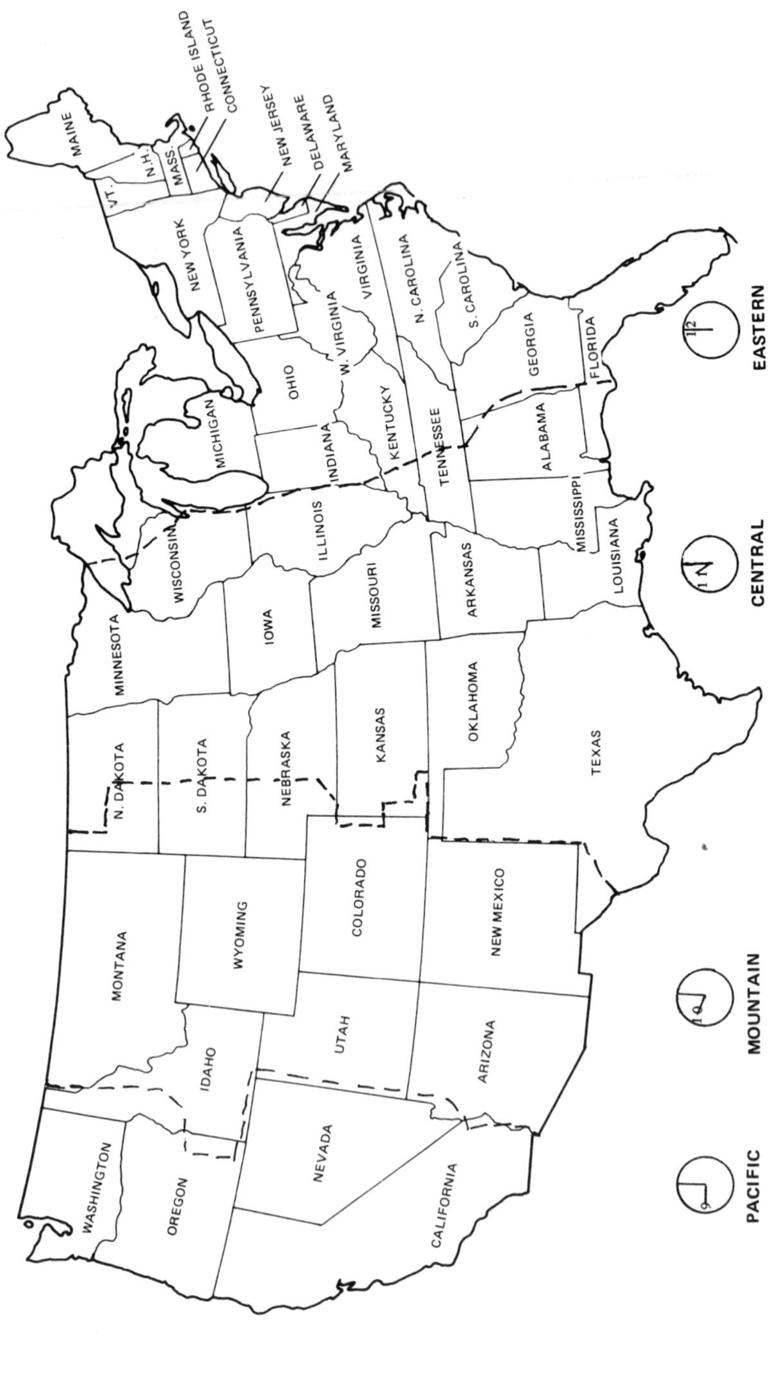

Figure 1-6. Map of the contiguous United States.

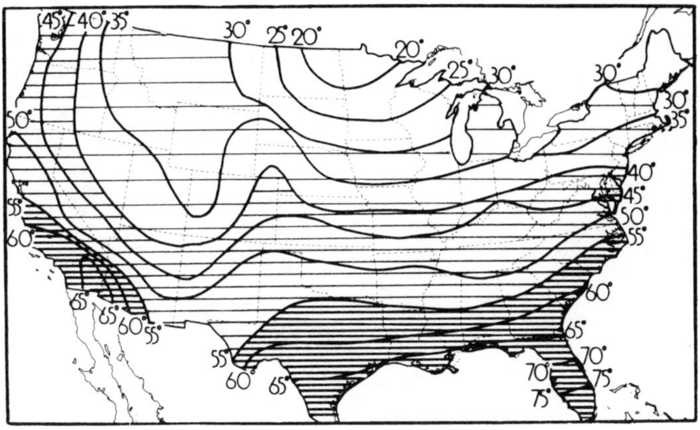

Normal daily maximum temperature, January 1–7.

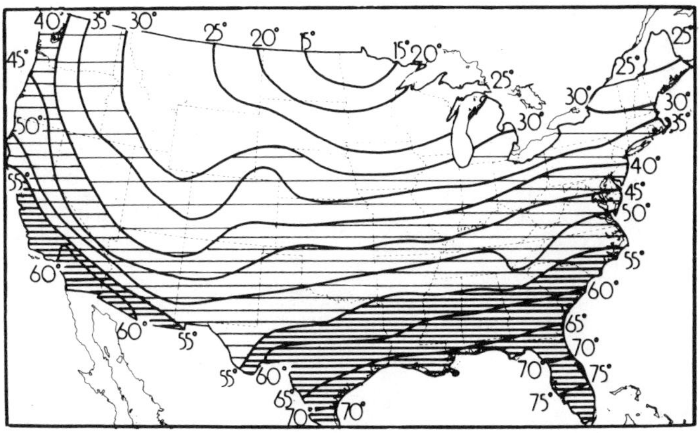

Normal daily maximum temperature, January 15–21.

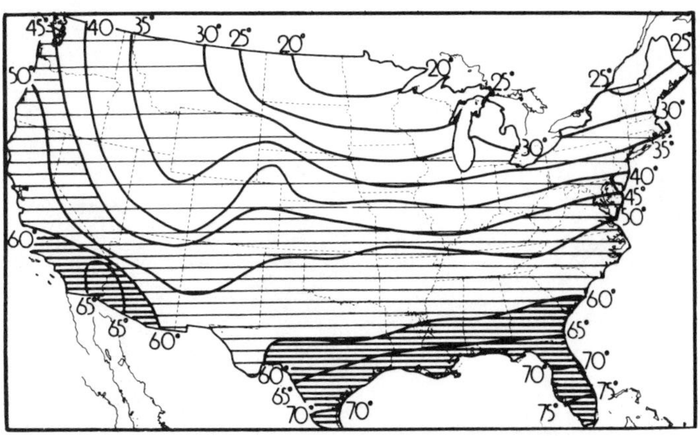

Normal daily maximum temperature, January 29–February 4.

Figure 1-7. Maximum temperatures, January (Visher, 1954).

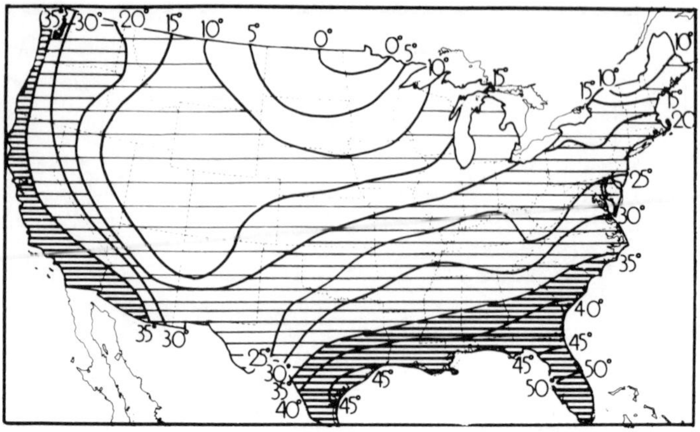

Normal daily minimum temperature, January 1–7.

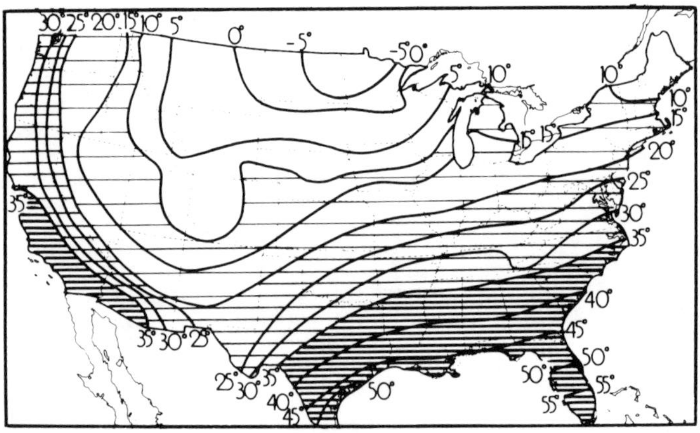

Normal daily minimum temperature, January 15–21.

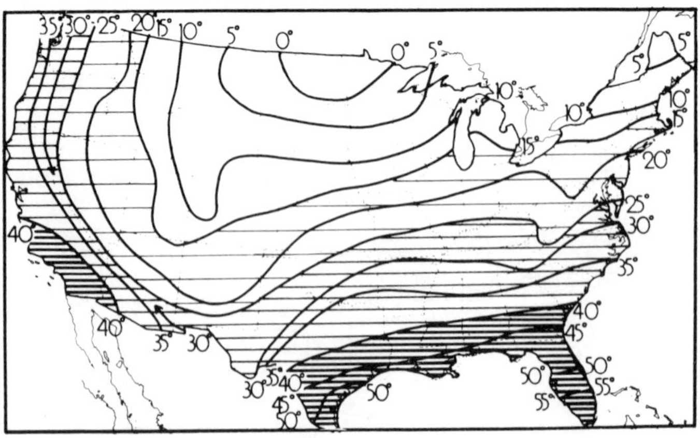

Normal daily minimum temperature, January 28–February 4.

Figure 1-8. Minimum temperatures, January (Visher, 1954).

necessary in winter the heat pump here is clearly operating with its best COP. Its relative economy will then depend a lot on its summertime utilization for air conditioning.

4. Southwest, California, Nevada, Arizona. Where daily temperature extremes are slightly chillier than those of the Deep South, heat pump economy will depend on summer air conditioning needs.
5. Northwest, Oregon, Washington. Where winter daily temperature extremes are closer to each other (30°F minimum, 45°F maximum) than anywhere else in the U.S., based on winter weather alone this area is probably the most propitious of all for a heat pump market.

Since the economy of the heat pump is also substantially influenced by its utility for air conditioning, we should judge the regional variation of this utility with similar charts on summer daily temperature extremes (Figures 1-9 and 1-10). In the North Central and New England regions there is a rather low air conditioning requirement, so that we have a combination of low winter effectiveness and low summer requirement. Neither will there be much of an air conditioning requirement for the entire Pacific Coast. This requirement is high for the remainder of the U.S., with a peak in Texas and the adjacent Midwest. Hence, the Deep South has the best combination of winter heat pump effectiveness and high summer air conditioning requirement.

ECONOMIC BASIS FOR HEAT PUMP MARKET PROSPECTS

Due to its handicaps compared to fuel burning, the heat pump was originally introduced in the southern U.S., where the winters are less severe. This area is still considered the best market. To establish a background for later discussion of technical trends aimed at relieving the heat pump's economic deficiencies, it is interesting to review its assessments compared to gas and oil heating in this region, and to electric resistance heating for other regions.

An economic analysis was made by Pickelsimer (1977) comparing the heat pump, gas-fired and oil-fired furnaces with electric air conditioning on the total-space, comfort-conditioning concept. Equipment efficiencies assumed are: gas-fired furnace, 45%; oil-fired furnace, 45%; electric air conditioning COP_c = 3. The heating value of natural gas was assumed to be 1000 Btu/ft^3. Figure 1-11 shows the result of an economic comparison of the heat pump and the gas-fired furnace. The retail cost of natural gas for residential use was about 24¢/therm* (equal to 82¢/kWh) and electricity

*1 Therm = 100,000 Btu = 29.3 kWh.

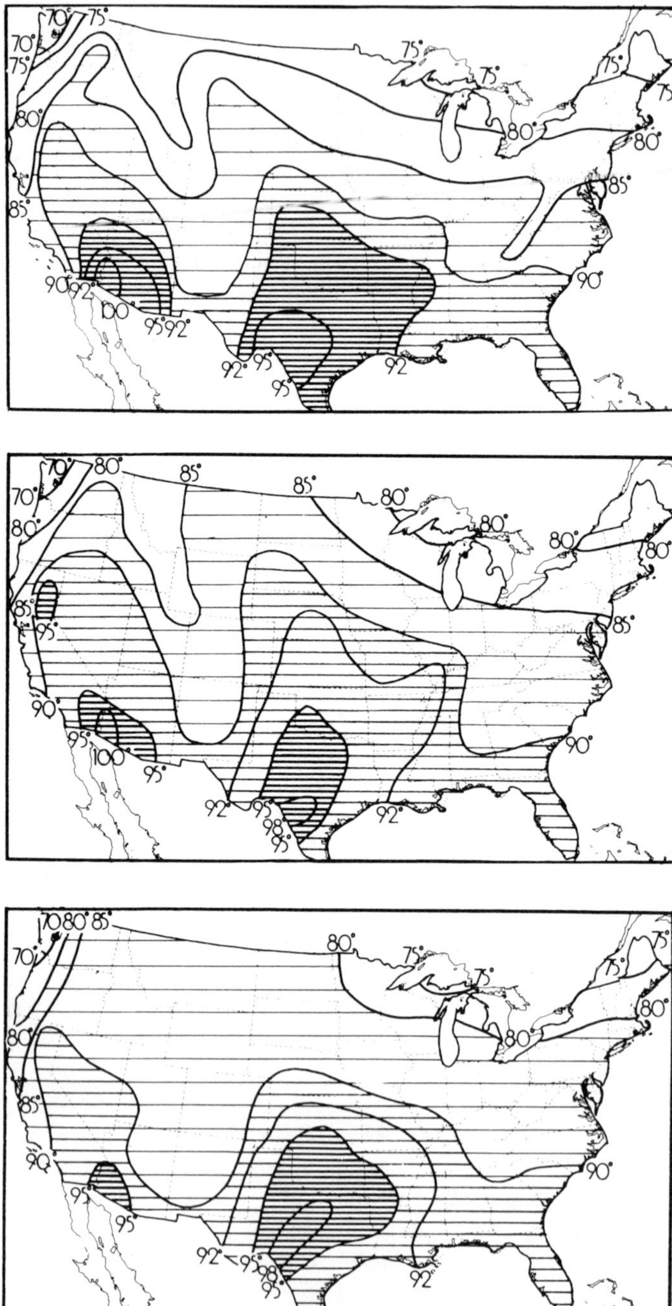

Normal daily maximum temperature, July 2–8.

Normal daily maximum temperature, July 16–22.

Normal daily maximum temperature, July 30–August 5

Figure 1-9. Maximum temperatures, July (Visher, 1954).

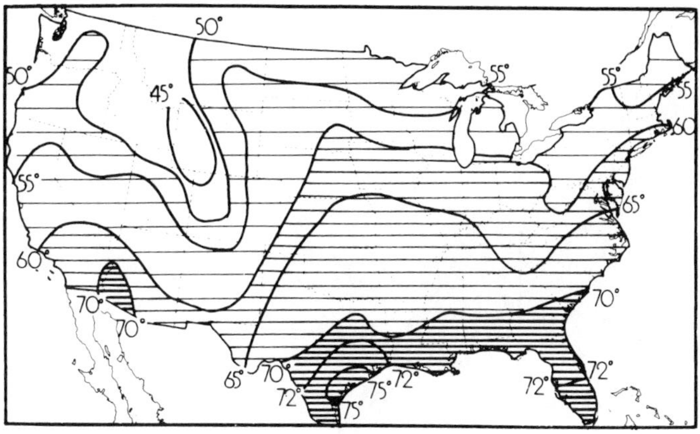

Normal daily minimum temperature, July 2–8.

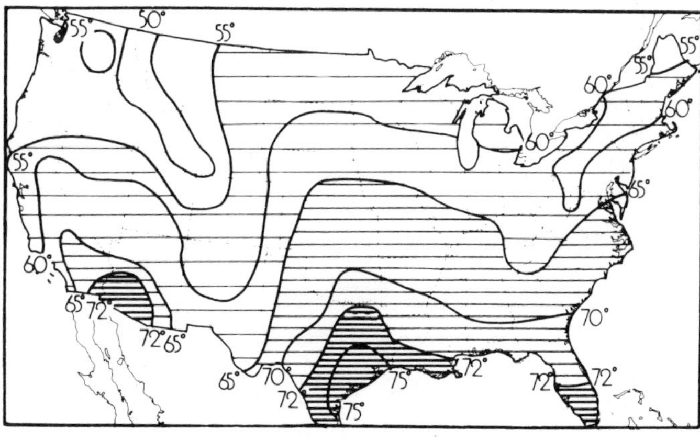

Normal daily minimum temperature, July 16–22.

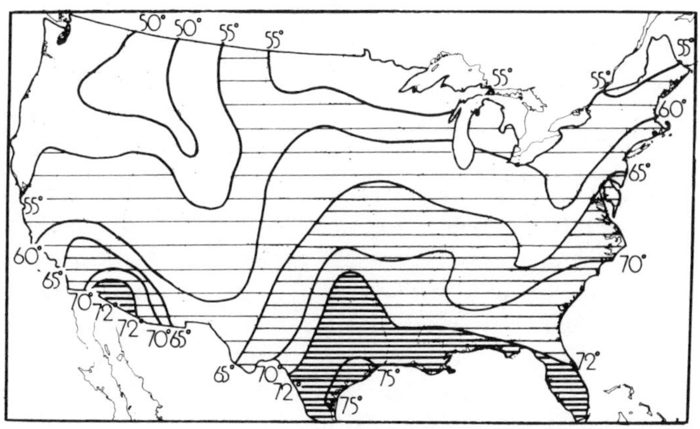

Normal daily minimum temperature, July 30–August 5.

Figure 1-10. Minimum temperatures, July (Visher, 1954).

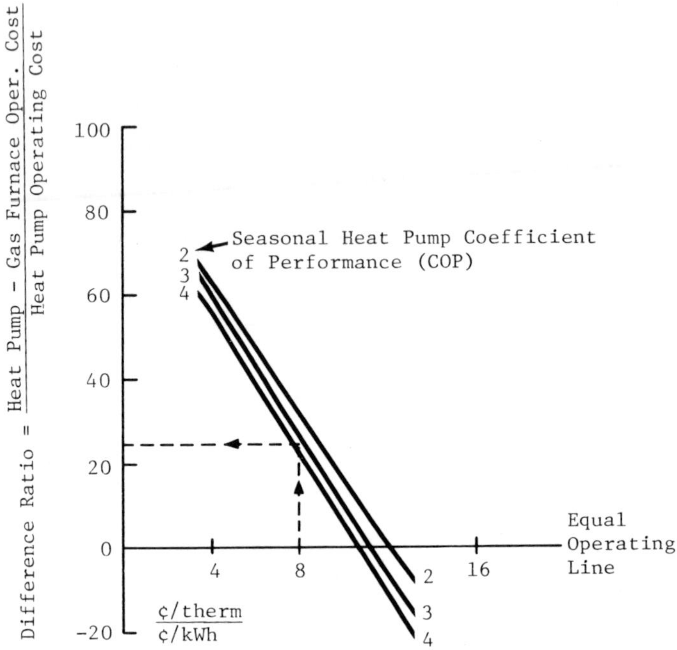

Notes:

1. Heating cost difference ratio vs unit energy cost ratio
2. Gas furnace efficiency = 45% with electric air conditioner in summer
3. Gas heating value - 1000 Btu/ft^3

Figure 1-11. Seasonal heating cost comparison for residential application–heat pump vs gas-fired furnace (Pickelsimer, 1977).

cost about 3¢/kWh giving a unit energy cost difference ratio of 8. At an energy cost difference ratio of 8 (see dashed lines in Figure 1-11), the heat pump is more expensive to operate by about 25% for a COP of 3. However, if electricity rates increase by 1.5 and gas rates triple by 1980, raising the energy cost ratio to 16, the heat pump will be more economical than the gas-fired furnace, as shown. Figure 1-12 shows the results of a comparison of the heat pump and an oil-fired furnace burning No. 2 fuel with a heating value of 140,000 Btu/gal. The current retail cost of No. 2

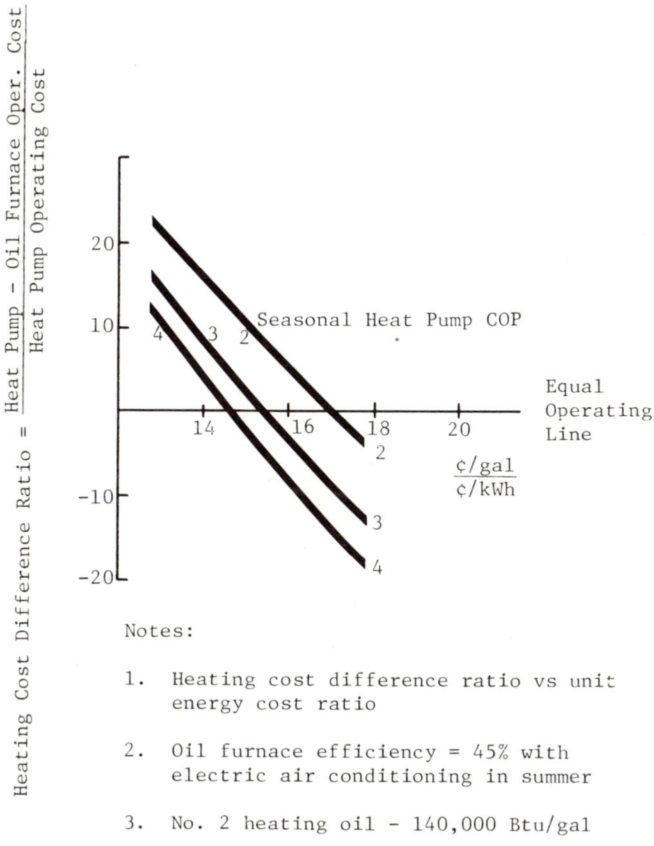

Figure 1-12. Seasonal heating cost comparison for residential application–heat pump vs oil-fired furnace (Pickelsimer, 1977).

oil is about 39¢/gal and 3¢/kWh for electricity, giving a unit energy cost ratio of 13. If the cost of electricity increases by 1.5 and the cost of No. 2 fuel oil triples by 1980, for an energy cost ratio of 26, the heat pump will be more economical on a total-space, comfort-conditioning concept. Figures 1-11 and 1-12 do not account for differences in capital cost, depreciation and maintenance, which must include air conditioning.

Table 1-1 compares the annual cost of operating a heat pump, a gas furnace with electric air conditioning and an oil furnace with electric air conditioning in a residential application. Depreciation has been figured on a 15-year equipment life and 40-year life for duct work. Capital cost has been figured on a 7% mortgage ratio of the total installed cost. Maintenance

Table 1-1. Heating Costs of a Single-Family Residence (1500 ft^3) in the Southeastern United States (1975 dollars) (Pickelsimer, 1977)

Item		Heat Pump ($)	Gas Furnace with Electric Air Conditioning ($)	Oil Furnace with Electric Air Conditioning ($)	Energy Costs		
					Electric	Gas	Fuel Oil
Operating Costs	1975	281	237	260	3¢/kWh	24¢/therm	40¢/gal
	1980	422	535	603	4.5	72	120
	1985	632	803	905	6.75	108	180
Depreciation		1900 @ 6.7% = 120	1400 @ 6.7% = 93	1400 @ 6.7% = 93			
		1200 @ 2.5% = 30	1200 @ 2.5% = 30	1200 @ 2.5% = 30			
					Annual Load:		
Capital Cost		3000 x 0.07 = 210	2600 x 0.07 = 182	2600 x 0.07 = 142	Heating -- 0.4 x 10^8 Btu/yr		
Maintenance		1100 x 0.04 = 44	200 x 0.02 = 4	800 x 0.04 = 4	Cooling – 0.4 x 10^8 Btu/yr		
			500 x 0.03 = 15	500 x 0.03 = 15			
Total Annual Cost	1975	685	561	604			
	1980	825	859	947			
	1985	1036	1127	1249			

has been figured on a percentage of the original equipment cost with a factor based on relative equipment complexity. Data were obtained by surveying heating and air conditioning contractors in the southeastern United States. Note that the heat pump system is more expensive to own and operate based on 1975 retail fuel prices. Pickelsimer (1979) has said that the typical 1975 fuel costs assumed were based on interviews with representatives of heating oil and natural gas suppliers. It was expected then that deregulation would raise fuel prices sharply. However, this deregulation still is not an accomplished fact, although the U.S. National Energy Act does relax controls and views eventual deregulation by the mid-1980s. With a 300% price increase in natural gas and fuel oil as compared to a 150% increase in electricity by 1980, the heat pump would be less expensive to own and operate nationwide than both oil and gas furnaces with electric air conditioning. At current prices of fuel and electricity, the heat pump capital cost differential and higher operating cost incur a present negative cash flow balance. The economic decision would be to wait a short while for heat pump technology to improve. Pickelsimer's review and analysis was part of the program sponsored by the U.S. Environmental Protection Agency (EPA) and is described further in Chapter 5.

The circumstances described are highly favorable to the heat pump. Fossil furnace efficiencies were assumed to be rather low and the capital cost interest charges are much higher in 1978. Annual costs for heating and air conditioning depend on the location. In the northern U.S., the mean temperature may be 10-20° lower than in the southern U.S. Therefore, heat pump efficiencies would be somewhat less, but fossil furnace efficiencies somewhat greater than in the south. The operating cost for the alternatives presented may change as much as 20%, depending on location, but trends would not change. In Alaska, for example, almost no homes are equipped with air conditioning. Based on the low efficiency of a heat pump at low ambient temperatures and the current fuel prices, a heat pump is not economical compared with a similar fossil-fueled system, when considering the heating mode only.

In a comprehensive study sponsored by the Electric Power Research Institute and the Association of Edison Illumination Companies, Blake and Gernert (1978) of Westinghouse monitored the operational characteristics of 120 heat pump systems installed for residential use in a wide variety of U.S. regions. The economic comparison was made exclusively with electric resistance heating. The energy inputs to 120 heat pump heating systems were measured directly and the energy supplied to the houses was indirectly measured at 43 of the test sites.

Energy use for heating may be related to the outside temperature through the concept of degree days, defined as:

$$DDH = 65^{\circ}F - T_{day}$$

where DDH = degree days heating for a particular day

T_{day} = average outside air temperature for that day, often taken as the average between the maximum and minimum temperatures.

The degree days for a heating season are then added for all days of the heating season in which T_{day} is less than 65°F. The winter season was taken as October 1-March 31, and measured energy use was related to the DDH at each test site.

By comparing the energy supplied by the heating systems to their energy input, they made estimates of the energy saved by heat pump systems relative to electric furnaces. The energy used by the heat pump heating systems was measured for each of the 120 houses. As part of the test procedure, many of these houses had the heat pump compressors switched off. Only resistance heating was used on alternate days. Thus, these heat pumps did not operate for the full heating season. To provide a base for common analysis, the energy used by each heat pump system was normalized by dividing by the heating degree days that the particular house had been exposed to while the heat pump compressor was not bypassed, then multiplying by the DDH for the entire season. The nonheating system electric energy used when the heat pump compressor was not bypassed was determined and reduced to an average hourly value.

The data grouped for each collaborating utility are presented in Table 1-2. The regions are ordered by heating degree days with minor shifts in order displayed between the 1975-1976 winter. Note that two regions were represented in the first winter only.

Column 3 represents the energy used per house for the heat pump heating system operating for the full season. Column 4 shows the energy use per degree day. This ratio is relatively constant over a range of degree days from 326-6666. Variations in the ratio are primarily a function of house construction, weather variables other than degree days, and lifestyle. Columns 5 and 6 give the minimum and maximum values of this ratio.

For those regions where data were collected through two winters, only those houses that had reliable data in both winters were used in these calculations, resulting in 105 out of the 120 being used for the first winter calculations. Two regions were not involved in the second winter, while other houses dropped out of the project, resulting in 96 houses under test for the second winter, with 85 being used for the calculations in Table 1-2.

It is clear from these results that for the more rigorous winters of the north central states and Rochester in the upper Northeast, the cold nights result in higher electric power per degree day. This reflects the lower operating COP in those low temperature extremes. Such a variation does

Table 1-2. Heat Pump Energy Use per Home, by Region (Blake and Gernert, 1978)

Degree Days[a] Heating (ddh)	Utility	kWh (HP)[b] (avg)	kWh(HP)/ddh (avg)	(min)	(max)
		First Winter, October 1975-March 1976			
326[c]	A Fla P&L	1187	3.64	0.86	6.61
1566	B Houston P&L	2537	1.62	0.78	2.62
2576	D Arkansas P&L	7255	1.90	1.51	2.42
2962	C Pacific G&E	7775	2.61	1.71	3.66
3196	E So Cal Ed	5561	1.74	1.15	3.27
4014	H P.S. New Mex	11159	2.78	1.40	4.08
4576	F Philadelphia Elec	12218	2.67	1.57	4.04
4602	G Con Ed N.Y.	12794	2.78	0.62	4.48
4603	I PSEG (NJ)	12428	2.70	1.78	4.78
5202	J Rochester G&E	18322	3.52	2.71	4.44
5817	K Consumers Pwr	12322	2.12	1.22	3.11
6666	L Northern States	20665	3.10	1.75	4.02
		Second Winter, October 1976-March 1977			
622[c]	A Fla P&L	1203	1.95	0.81	3.83
2379	B Houston L&P	4377	1.84	0.99	2.90
2782	C Pacific G&E	6788	2.44	1.66	3.38
2927	E So Cal Ed	5913	2.02	1.46	2.90
3313	D Arkansas P&L	7255	2.19	1.48	2.88
4620	H P.S. New Mex	12428	2.69	1.37	4.49
5269	G Con Ed N.Y.	21129	4.01	2.26	6.11
6536	J Rochester G&E	22288	3.41	3.00	4.19
6837	K Consumer Pwr	12101	1.77	0.60	2.59
7565	L Northern States	22090	2.92	1.47	3.82

[a]ddh is the six-month total degree days heating.

[b]kWh(HP) is the energy used by the heat pump heating system, including indoor fan.

[c]Because of the few hours below 65°F in this region, kWh was based only on those hours where the temperature was below 65°F. This is equivalent to a shorter heating season.

not occur for fuel burning as the heat source. Here, the energy required can be linearly related to the degree days and this linearity was the historical reason for adopting the concept to estimate charges.

Wide variations in energy use per degree day from house to house can be expected. Average energy use per house per degree day shows considerably less regional variation than the house-to-house variation within regions. Since the number of samples is small, the data are only rough

Table 1-3. Estimated Average Energy Saved per Test House by Region–Heat Pump vs Electric Furnace (Blake and Gernert, 1978)

October 1975–March 1976				October 1976–March 1977			
Region	ddh	SPF[a]	kWh Saved[b] per Test House	Region	ddh	SPF[a]	kWh Saved[b] per Test House
A	326	2.25	1484	A	662	2.21	1456
B	1566	2.07	2715	B	2379	1.94	4114
D	2576	1.91	6602	C	2782	1.88	5973
C	2962	1.86	6687	E	2927	1.86	5085
E	3196	1.82	4560	D	3313	1.80	5804
H	4014	1.70	7811	H	4620	1.61	7581
F	4576	1.61	7453				
G	4602	1.61	7804	G	5269	1.51	10776
I	4603	1.61	7581				
J	5202	1.52	9527	J	6536	1.32	7132
K	5817	1.43	5298	K	6837	1.27	3267
L	6666	1.30	6200	L	7565	1.17	3755

[a]SPF estimated by $SPF = 2.3 - 0.15 \times 10^{-3} \times ddh$.
[b]kWh saved estimated by $kWh\ saved = kWh(HP) \times (SPF-1)$.

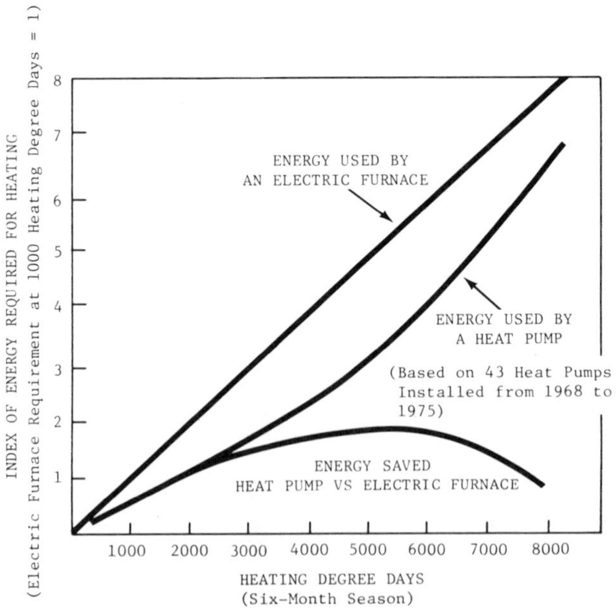

Figure 1-13. Energy saved by using a heat pump compared to an electric furnace (Blake and Gernert, 1978).

estimates of the average energy use in each region; however, a good estimate of the average energy use is provided and is valid for utility purposes. The energy saved by using a heat pump rather than electric furnace (Table 1-3) is considerable in all regions. The percentage of energy saved is greatest in the warmer regions, while the total energy saved tends to reach a maximum for regions with about 5000 DDH seasons, as shown in Figure 1-13. The question of how economical a heat pump investment would be requires information on capital and maintenance costs for the system. This question was left open. Market prospects for heat pumps that reflect these current technical realities and lead to strong variation in regional acceptance are described in Chapter 5.

CHAPTER 2
CURRENT ELECTRIC HEAT PUMP TECHNOLOGY

INTRODUCTION

This chapter mainly concerns air-to-air heat pumps operating as both heat pumps and air conditioners. They are still the predominant type and attract most of the attention for improvement in effectiveness and reliability for use in residential and light commercial buildings.

Emphasis is on unitary equipment, i.e., equipment manufactured in one or more complete assemblies designed to be operated as a unit. Applied or "built-up" heat pump systems are not considered. Because these heat pumps are assembled from components at the building site and because of the size, i.e., refrigeration tonnage, and large capital investment usually involved in single installations, large built-up heat pump systems have frequently offered more opportunities for innovative engineering and efficiency improvement than unitary equipment. Many of the technical approaches that have been suggested for improved efficiency of unitary heat pumps are already being practiced in built-up heat pump systems. On the other hand, unitary equipment presents a specific challenge for the industry to improve equipment performance and reliability at a cost that competes with available and traditional alternatives.

In Chapter 1, the coefficient of heat pump performance as a conveyor of thermal energy from a cold to a hot reservoir was related to its thermodynamic parameters of energy conveyed and required work under ideal conditions.

No practical unit will operate at the thermodynamic limit. Suppose that, because of friction and other kinds of imperfections in the design, the apparatus operates at some fraction, f, of the theoretical efficiency. We can take f as a measure of the unit's effectiveness. Then the work available W will be:

$$W = Q_2 \times f \times N$$

where N = $(T_2-T_1)/T_2$. T_2–hot reservoir temperature; T_1–cold reservoir temperature

(1-f) = overall coefficient of friction to represent energy dissipated by nonideal operation of the engine

W = work available after heat, Q_2, is used to drive the engine with exhaustion of heat Q_1 at temperature T_1

Ideal coefficient performance would then be expressed as:

$$COP = Q_2/(Q_2-Q_1) = \frac{T_2}{T_2-T_1} = 1/N$$

If the cycle were reversed to produce the effect of conveying heat from the cold to the hot reservoir, the work required will be amplified by the same functional loss according to the factor 1/f, that is:

$$W = Q_2 N/f$$

When our system operates as an engine, we get less work than theory calls for, the difference being 1-f times theoretical work, Q_2N. If the system operates as a pump, work is required in excess of theoretical requirements, the difference being (1-f)/f. In this case, the quantity of work dissipated (1-f), Q_2N, or (1-f)/f, Q_2, can normally be lost. However, for heating, one can arrange that this dissipation be delivered to the hot reservoir so that total heat delivered is:

$$Q_2' = \frac{Wf}{N} + (1-F)W$$

the coefficient of performance is then:

$$\frac{Q_2'}{W} = \frac{f}{N} + 1-f$$

instead of 1/N in the ideal case.

From the viewpoint of heat economy, any work cycle will be better than a resistance heater for which f = 0 and COP = 1. The salient questions are, of course: How much better? and, When is it justified?

Hence, the performance of a practical heat pump depends primarily on three factors: (1) the effectiveness of the cycle used; (2) the inverse of the temperature interval; and (3) the absolute temperature of the hot side. If a thermodynamically perfect cycle were available, only the product of (2) and (3) would have to be considered. Some numerical comparisons based on maintaining a hot side temperature of 70°F (21°C) are given in Figure

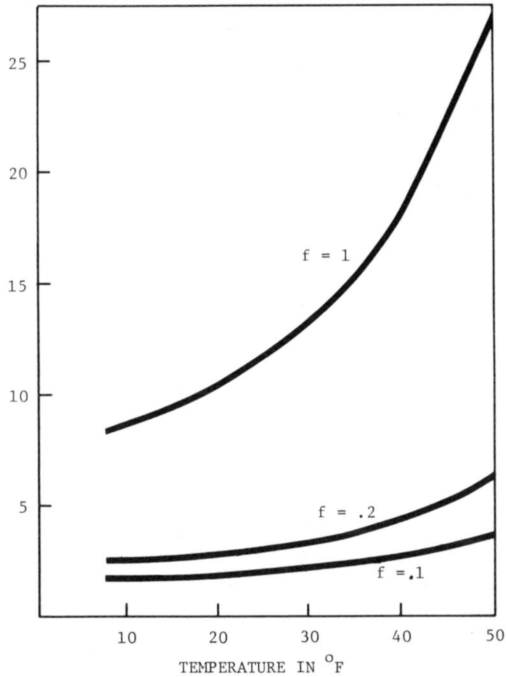

Figure 2-1a. Theoretical COP as a function of partial ideal efficiency and temperature.

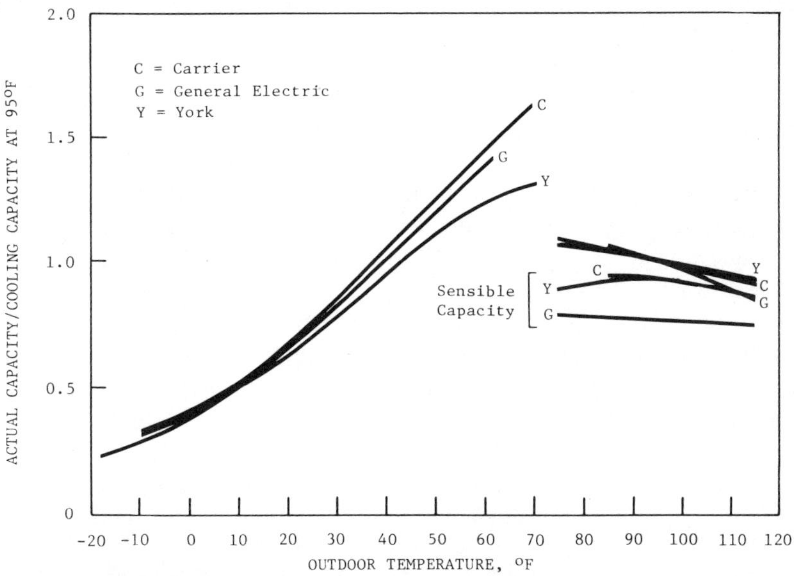

Figure 2-1b. Normalized heating and cooling capacity of typical heat pumps (based on total cooling capacity at 95°F outdoors) (Gordian Associates, 1978).

2-1a. Cold source temperatures ranging from 10°F (-12°C) to 50°F (10°C) were assumed, and coefficients of performance corresponding to three different values of cycle effectiveness—theoretical, f = 1; practical device with minimal extraneous losses, f = 0.2; and the apparent current level of f = 0.1—were considered. A device capability of 0.5 represents a rather optimistic estimate of what might be accomplished with good mechanical design and the actual properties of the available working fluids. Real machines will suffer from additional losses—thermal drops across heat transfer surfaces, power consumed by fans and control circuits, the energy needed to operate auxiliary heaters, etc. Further, in the end, we will have to consider the energy required to operate defrost cycles when needed. These problems are discussed later. Meanwhile, for purposes of comparison, the lowest curves of Figure 2-1a represent the range of currently feasible COP values calculated, assuming a cycle effectiveness, f, of 0.2 and 0.1.

CAPACITY AND EFFECTIVENESS

Because rate of heat transfer is reduced as ambient temperatures reach extremes for exchangers in both heating and cooling modes, the heating and cooling capacity and coefficient of performance of the electrically driven Rankine cycle heat pump vary in the worst way as functions of the outdoor temperature. Figure 2-1 shows the temperature dependence of heat pump heating and cooling capacity, normalized according to rated cooling capacity at 95°F outdoor dry bulb temperature for air-source equipment of three different manufacturers.* The heating *capacity* of the heat pump decreases with the ambient temperature; the cooling capacity decreases with an increase in the ambient temperature, which is the reverse of the heating and cooling demand of most buildings. Figure 2-2 shows the variation with ambient temperature of the heating and cooling COP. Again, the heat pump is least effective when the heating or cooling demand is greatest.

In addition to the drop in COP due to the fall in the basic thermodynamic driving parameters (hot reservoir absolute temperature divided by temperature difference between hot and cold reservoirs), as explained in Chapter 1 and Figure 2-1a, there are two other reasons for the decrease in heat pump capacity and COP with ambient temperature. First, the defrost cycle is frequently necessary for temperatures between 32°F and

*The data represent a General Electric 10-ton single package heat pump, a Carrier 4-ton split system heat pump and a York 2-ton split system heat pump: 1 ton = 11,950 Btu/hr (837 cal/sec). A capacity of 1 ton is equivalent to the capability of freezing or melting 1 ton of ice in 24 hr (10.5 g/sec). Sensible capacity excludes energy extraction serving only to condense moisture.

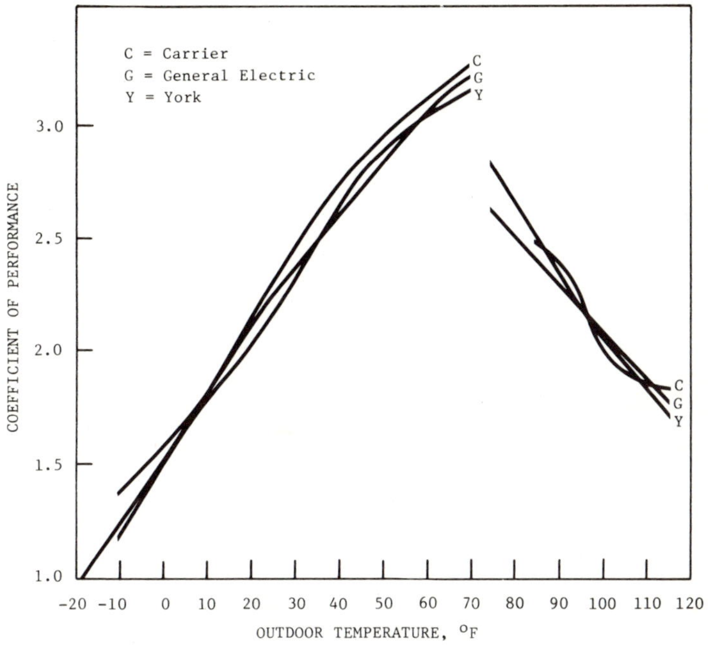

Figure 2-2. Heat pump heating and cooling COP (Gordian Associates, 1978).

41°F (0°C and 5°C). The extent of this influence depends on the relative humidity variably experienced in the region and with time each day in the heating season. If humidity is high enough, depending on the specific defrost cycle control routine of the manufactured unit, operational COP may fall below 1 because of the *extra* supplementary heat required to compensate for the heat extracted from the hot reservoir during the cycle.

The second major factor contributing to low performance in a real heat pump cycle is the need to use significant temperature gradients to provide heat transfer from the source to the evaporator coil and from the condenser to the heat sink. In general, the condenser must operate 10-20°F hotter than the medium heated, and there will also be a corresponding temperature drop to transfer heat from the source to the evaporator. The lower the ambient temperature of the air entering the outdoor coil (evaporator) (Figure 2-3), the lower the evaporator temperature has to be to maintain a sufficient temperature difference for heat transfer to the vaporizing fluid. As the evaporator temperature drops, the evaporator equilibrium pressure decreases, but the compression ratio increases. The effect on the refrigerant mass flow is thus twofold: (1) the inlet gas density decreases due to a lower pressure (at roughly constant superheat); and

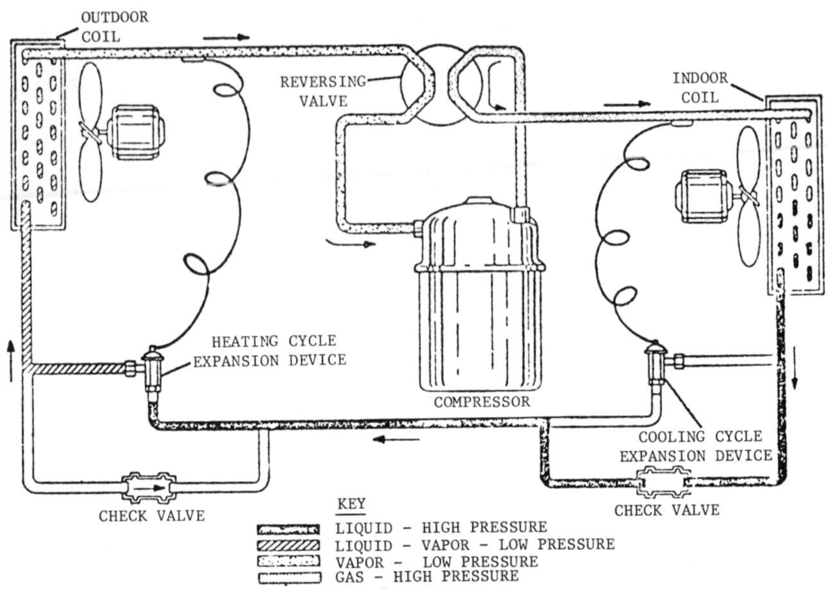

Figure 2-3. Heat pump refrigerant loop schematic, heating mode (Westinghouse Electric Corporation).

(2) the volumetric efficiency decreases due to the higher compression ratio. The result is a lower mass flow of refrigerant and an increase in the required specific work of compression.

In the cooling mode shown for the Westinghouse systems (Figure 2-4), cooling capacity and COP decrease as the outdoor temperature rises. The outdoor coil, now the condenser, must be maintained at a temperature 10–15°F higher than the entering ambient air for adequate heat transfer. As the required condensing temperature rises with higher ambient temperature, the compression ratio increases, increasing the work of compression and decreasing the volumetric efficiency. The attendant decrease in cooling capacity and COP is somewhat offset by the higher suction gas density.

The latent heat removal or dehumidification capacity of the heat pump also varies with temperature. When the evaporator (or indoor coil) temperature is low (45-50°F), there exists a sufficient vapor pressure difference between the warmer, more humid entering air and the cooler stagnant vapor film over the condensed liquid water at the coil surface to cause rapid mass transfer, the heat of water vapor condensation being absorbed by the evaporating refrigerant along with the sensible heat of the air being cooled. With increasing coil temperature outdoors and evaporator temperature

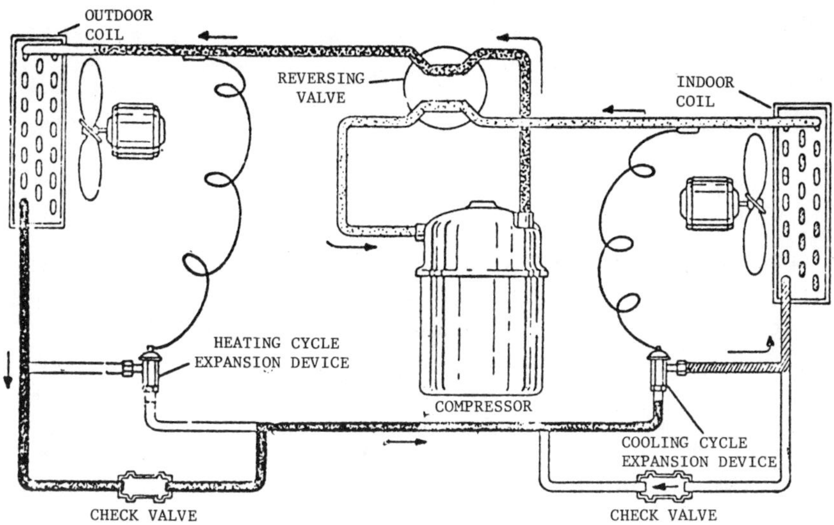

Figure 2-4. Heat pump refrigerant loop schematic, cooling mode (Westinghouse Electric Corporation).

indoors, the bulk air-to-surface film water vapor pressure differential at the indoor coil decreases and the dehumidification rate drops. That is, the latent heat removal or cooling capacity of the heat pump decreases with increasing outdoor temperature. On the other hand, at a given outdoor entering air temperature, the total capacity increases with indoor wet bulb temperature due to increased condensation rather than sensible heat transfer. See the total and sensible cooling capacity curves in Figure 2-1. When the coil temperature approaches entering air, wet bulb temperature, moisture removal capability ceases. This is of concern in ensuring cooling comfort and may impose an upper limit on the indoor heat exchanger size (and thus a limit on one means of efficiency and capacity improvement).

CONTROLS

Unitary heat pumps employ a number of controls for operational mode activation and termination, motor-compressor protection, refrigerant flow regulation, defrost initiation and termination, indoor and outdoor temperature sensing and, in recent models, user's and servicer's diagnostics. Developments in solid-state sensing and switching devices have substantially increased the speed of response and reliability of motor protection devices. Microprocessor technology plays a growing role in small HVAC system control schemes, enabling the monitoring, analysis and regulation of a larger number of parameters. For example, the York Champion heat pump uses a solid-state logic module, which monitors the following:

- air temperature entering outdoor coil (thermistor);
- compressor discharge line temperature (thermistor);
- liquid line temperature (thermistor); and
- pressure differential across outdoor coil (differential pressure switch).

Other features provided include an adjustable balance point control, with settings in the 23-43°F range, which prevents supplemental resistance heat operation above the balance point (in lieu of a field-installed outdoor thermostat). Plug-in contacts also allow interconnection with a servicer's diagnostic device to facilitate corrective maintenance. The York control system is described later in this chapter for normal operational modes.

In addition to motor protection and other devices having an obvious effect on equipment reliability, several control elements (valves) influence heat pump capacity. Therefore, some characteristics of capillaries, thermostatic expansions valves and subcooling valves will be discussed below. Further, some of the control features of individual manufacture units will be indicated in the section on available heat pump systems.

A capillary tube (a small-diameter coiled tube in the liquid line leading to the coil serving as an evaporator) produces a pressure (and corresponding temperature) drop in the entering fluid. Because the capillary is a device that presents a fixed constriction to refrigerant flow, it is unable to significantly adjust mass flow to changing upstream or downstream pressure conditions. If a capillary on the outdoor unit is sized to give 10-15°F superheat at moderate evaporator temperature, it will allow just enough liquid to pass at low evaporator temperature to improve compressor cooling and oil return.

A thermostatic expansion valve, on the other hand, varies the flow constriction (pressure drop) to maintain a constant amount of superheat in the evaporator effluent vapor. Thermostatic expansion valves used to be available only in the larger split-system heat pump sizes. Several manufacturers now use thermostatic expansion valves for both the outdoor and indoor coils, or for the outdoor coil only in the smaller, residential-size equipment. Thermostatic expansion valves provide better evaporator flow control; however, at low mass flows and evaporator temperatures, they unduly restrict refrigerant flow if set for constant superheat. The excess liquid refrigerant not admitted into the evaporator then collects in the condenser, restricting the coil area available for condensation and increasing condenser pressure. Thermostatic expansion valves use a metal diaphragm and a sensing fluid charge, which may be a source of failure.

An additional refrigerant control sometimes incorporated into the heat pump design is a pressure-activated subcooling control valve. A pressurized bulb senses the condenser effluent temperature with respect to its set point of 10-15°F from saturation temperature; hence, the extent of liquid subcooling. This concept is used in the Westinghouse system.

COMPREHENSIVE MICROPROCESSOR HEAT PUMP CONTROL

The York division of Borg-Warner has incorporated solid-state logic in the control of its line of heat pump systems. Refrigerant flow through the physical elements of their systems is illustrated in Figure 2-5, and individually for cooling and heating in Figure 2-6.

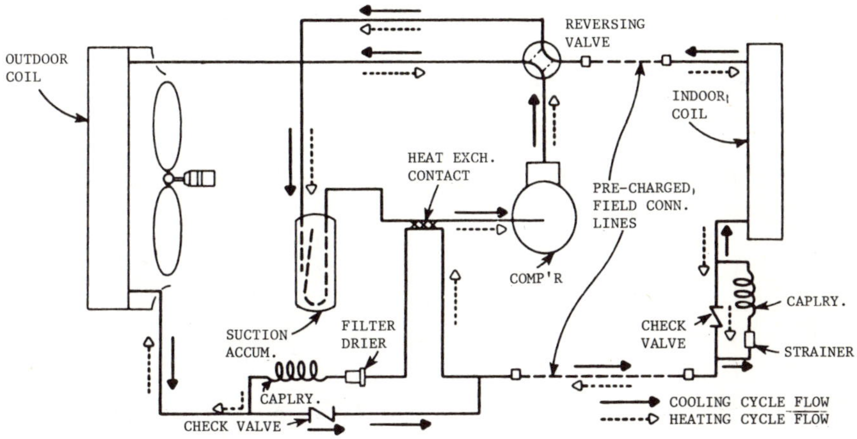

Figure 2-5. York split-system heat pump flow diagram (York Division of Borg-Warner).

Cooling Operation

For space cooling (Figure 2-6a), the compressor discharge gas flows through the four-way reversing valve to the outdoor coil, where it condenses to a liquid. Subcooled liquid leaves the outdoor coil and flows through a check valve to the indoor unit where it passes through a strainer and the capillary tube to the indoor coil.

The capillary tube restricts the flow as the refrigerant expands from a high to a low pressure. Evaporation takes place in the indoor coil, and slightly superheated gas flows through the vapor line to the outdoor unit. The gas flows through a parallel port in the four-way reversing valve to

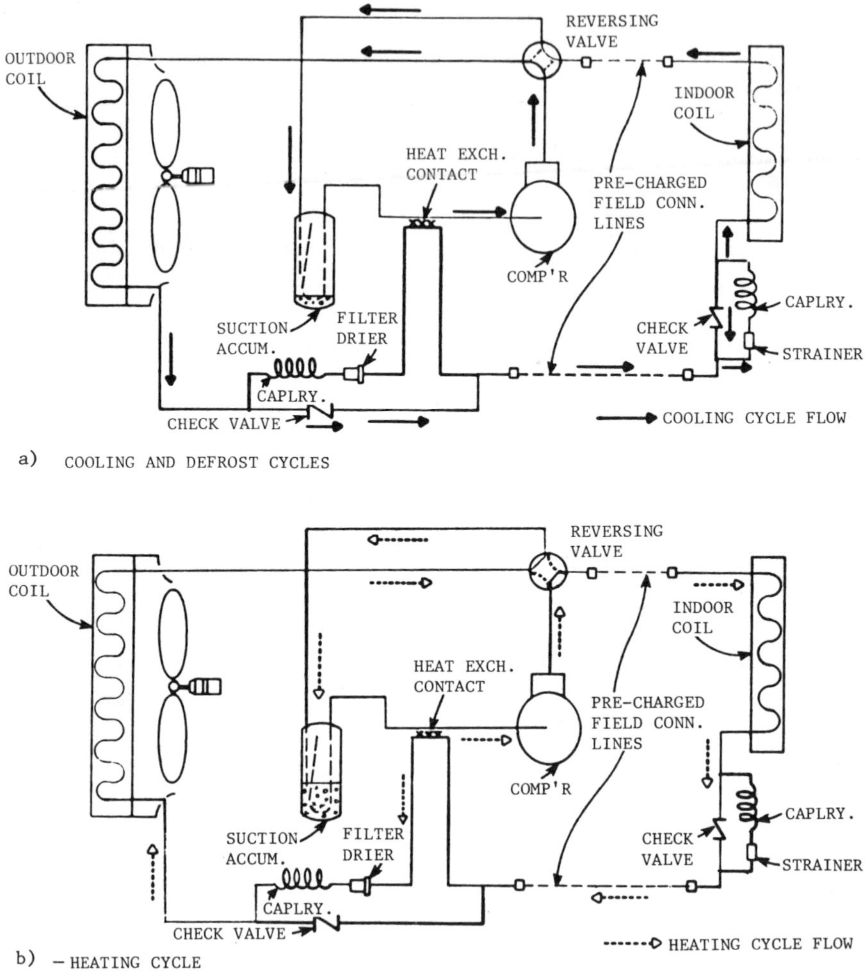

Figure 2-6. York split-system heat pump individual flow diagram (York Division of Borg-Warner).

the suction line accumulator and the suction line heat exchanger to the compressor suction to complete the cycle. Essentially, no liquid flows through the liquid side of the suction line heat exchanger during the cooling cycle, so no superheat is added to the suction gas. For cooling operation, the accumulator does not store refrigerant liquid, so heat exchange is not needed nor even desirable, as it would tend to overheat the gas.

Defrost, which is systemically similar to cooling in the operation of physical components, is described in the section entitled Coil Freezing and Defrost, page 44.

Heating Operation

For heating (Figure 2-6b), the four-way reversing valve is deenergized so that compressor discharge gas flows through the vapor line to the indoor unit. As the vapor passes through the indoor coil, it condenses to a liquid and becomes subcooled. It then flows through a check valve, bypassing the capillary tube, to the outdoor unit.

The check valve in the outdoor unit forces all the liquid refrigerant through the suction line heat exchanger, the filter-drier and the capillary tube. The capillary tube restricts the flow as the refrigerant expands from a high to a low pressure. As the low-pressure refrigerant flows through the outdoor coil, it evaporates and flows as a vapor through the four-way valve to the suction line accumulator, suction line heat exchanger and suction line to the compressor to complete the cycle.

When the outside air temperature is below 50–55°F, less refrigerant charge is required in the system than is needed for cooling operation. Therefore, the excess refrigerant in the system ends up in the suction line accumulator because the capillary tube tends to keep the indoor coil, operating as a condenser, fairly well drained. As a consequence, the accumulator may be empty of refrigerant liquid at 55°F outside air, and it may be nearly half full when operating on the heating cycle at -10°F. In other words, the system charge in the heat exchangers adjusts to satisfy the load and the excess ends up in the accumulator.

Since the suction line accumulator is a liquid separation device, oil, as well as refrigerant liquid, will separate from the gas, with liquid lying in the lower portion of the accumulator. Oil must be returned to the compressor to replenish its supply, so a small metering orifice is located at the bottom of a U-tube inside the shell. The U-tube is opened at one end near the top of the shell so it can flow gas that has been separated from the liquid. The gravity head of the liquid in the shell forces small quantities of refrigerant liquid and oil into the suction tube by way of the metering orifice, where it mixes with gas and returns to the compressor. The suction line heat exchanger is provided to evaporate these droplets of refrigerant liquid to avoid crankcase dilution of the oil.

Balance Point Determination and Setting

The balance point of a heat pump is the lowest temperature at which the refrigeration cycle can heat the building, unaided by supplementary heat.

The balance point depends on:

- outdoor design temperature
- building heat loss at design temperature
- unit capacity

With this information, the balance point can be determined for any installation. After the balance point for a particular installation has been determined, the set point or wire setting should be recorded and passed on to the installer. The following example illustrates the procedure to use. Assume:

> +10°F Design Outdoor Temperature
> 30,000 Btu/hr Building Heat Loss @ 10°F
> CHP30 System has been selected (nominal capacity 30,000 Btu/hr @ 45°F)

For this explanation, refer to Figure 2-7, which presents the heat pump capacity curves for all units in a sequence CHP024 through CHP060.

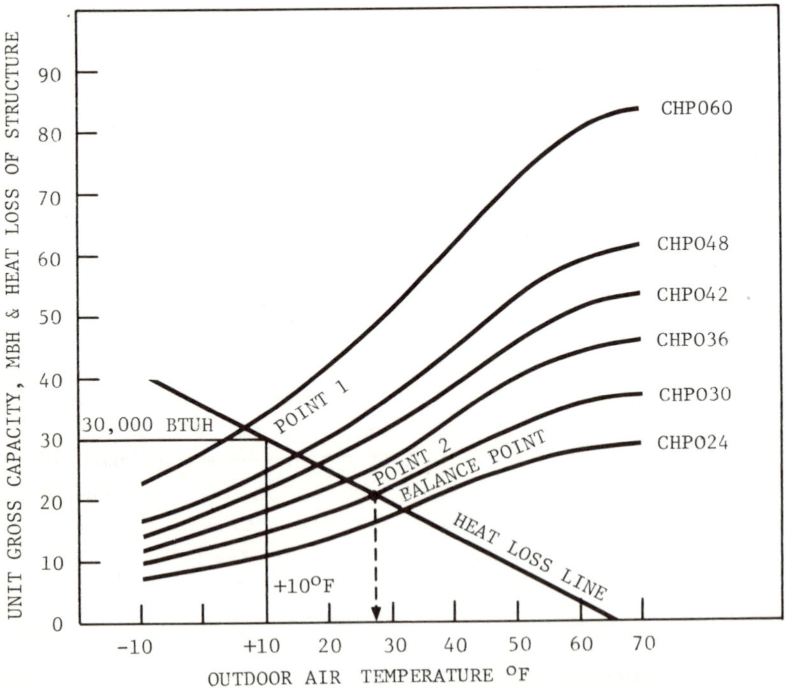

Figure 2-7. Heat pump capacity dependence on outdoor temperature for York units CHP024 through CHP060 (York Division of Borg-Warner).

1. Along the left side of the chart locate 30 1 (MBH) and draw a horizontal line.
2. Along the bottom of the chart locate +10°F outdoor temperature and draw a vertical line.
3. The intersection of these two lines, labeled point 1, is the building heat loss at design temperature.
4. Draw a straight line from point 1 to the 65°F outdoor temperature. This is called heat loss line. Any point on this line provides the building heat loss at the corresponding outdoor temperature. The heat loss is drawn to 65°F, since the concept of "degree days" assumes that at 65°F the internal heat sources will produce sufficient heat to balance the heat loss.
5. Point 2, the balance point, is at the intersection of the heat loss line and the heat pump curve which, in this example, is CHP030. At the balance point, the heat pump will supply exactly the heat needed to counteract the heat loss. In this example, the balance point is 27.5°F. At this balance point, the unit output is 20.5 MBH. At outdoor temperatures below 27.5°F, supplemental heat is required as the heat loss is greater than the capacity of the heat pump.

Control by York's Yorkguard Solid-State Logic Modules

The solid-state module, functioning as control microprocessor for all of York's heat pump systems, is shown in Figure 2-8. Appearance and spatial arrangement of the physical components governed by the module are shown in Figure 2-9. The compressor envelope, shown in the low right quadrant of the picture, is typically about 30 cm high and 15 cm in diameter. Descriptions of the major control functions exercised by the logic model are presented in Table 2-1. The module exercises a sequence of the system's operational steps according to the basic conditions of heating and cooling as follows:

With Power to Unit and Thermostat in Cooling Position

1. Reversing valve is energized through thermostat system switch to position refrigerant circuit for cooling operation.

2. If fan switch is in "ON" position, a circuit is made through blower relay providing continuous blower operation.

3. Solid-state Yorkguard starts time-delay period. After 5 minutes, unit is ready for operation.

4. When thermostat cooling contact closes, a circuit is made through the Yorkguard to energize contactor and start the unit. With fan switch in "AUTO" position, a circuit is made from thermostat cooling contact through blower relay to provide blower operation.

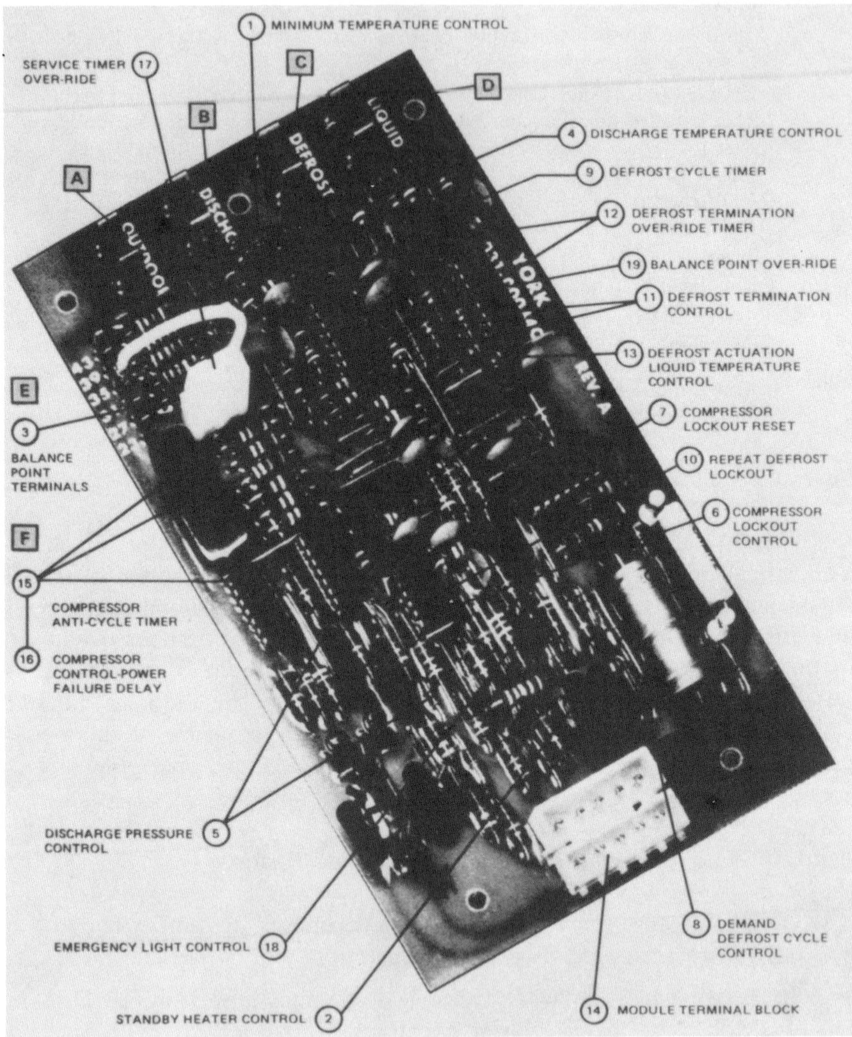

Figure 2-8. Yorkguard solid-state logic module (York Division of Borg-Warner).

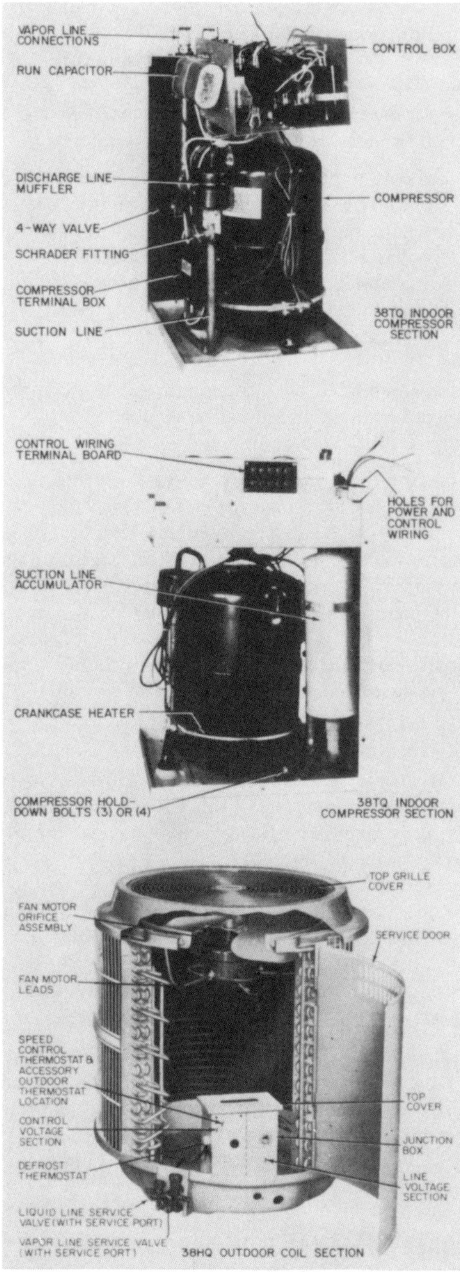

Figure 2-9. Outdoor compressor compartment components for York heat pump systems (York Division of Borg-Warner).

Table 2-1. Major Functions of Yorkguard Solid-State Control Logic Module (York Division of Borg-Warner)

A. "Outdoor" Sensor–connected to a thermistor located in the entering air stream of the outdoor coil to measure outdoor air temperature.

1. Provides low ambient temperature control to prevent the operation of the heat pump below -10°F. This prevents operation of the heat pump at high compression ratios, where the stresses on the compressor become severe.
2. Automatically activates the standby heaters if used under control of the room thermostat when outdoor air temperature is below -10°F.
3. Provides monitoring of outdoor air temperature to permit the "Adjustable Outdoor Balance Point Thermostat" to control the sequencing of electric heat.

B. "Discharge" Sensor–connected to a thermistor on the compressor discharge line and electrically linked to high-pressure cutout device, which responds to refrigerant discharge pressure and temperature.

4. Provides compressor high discharge temperature protection.
5. Provides compressor high-pressure protection.
6. Compressor lockout will occur for discharge temperature exceeding 275°F or discharge pressure exceeding 385 psig. Protects against loss of refrigerant charge or low air flow across indoor or outdoor coils.
7. Permits reset of compressor circuit to normal operation following lockout.

C. "Defrost" Terminals–electrically connected to a pressure differential switch, which measures the air pressure change across the outdoor coil.

8. Provides defrost on demand, which saves energy by initiating a defrost cycle only when needed.
9. Prevents nuisance defrosts caused by gusty winds. Defrost can only be initiated if the air pressure switch contacts have been closed for an accumulated interval of 12 seconds more than they have been open and the liquid line temperature is 39°F or lower (refer to no. 13).
10. Locks out the heat pump if it attempts to defrost within 5 minutes of the termination of the last one as this is an indication of a malfunction caused by deep snowdrifts, debris blocking the outdoor coil or a faulty defrost switch.

D. "Liquid" Sensor–connected to a thermistor on the liquid line and measures liquid refrigerant temperature.

11. Provides termination of defrost when the liquid line temperature exceeds 75°F.
12. Provides defrost cycle override. In the event that the wind prevents the liquid temperature from attaining the normal defrost termination temperature, a timed override will terminate the defrost cycle.
13. Prevents defrost initiation during mild weather. Liquid temperature must be 39°F or lower to initiate defrosting.

E. Adjustable Balance Point Control–prevents the operation of electric heat above the balance point. This prevents supplemental heat operation when the heat pump can satisfy the building load. The balance point is adjustable to match the climate and can be set at 43°F, 39°F, 35°F, 31°F, 27°F and 23°F. Removes the need to purchase and field install an accessory outdoor thermostat.

Table 2-1, continued

F. "Five Minute Timer."

14. Provides a 5-minute anticycle delay, which prevents restarting in less than 5 minutes after last shutdown. Unit will have time to equalize pressures before restarting. Thermostat chatter will not affect unit.
15. Following a power failure, delays start of the compressor for 5 minutes, allowing sufficient time for the line voltage to be restored to normal.

Others

16. Plug-in electrical connector interfaces module with system components and may be used to connect service analyzer.
17. Service connection timer override.
18. Compressor lockout feedback to room thermostat indicator light.
19. Emergency heat override of balance point controller.

5. Unit will cycle in response to thermostat signal to provide cooling as needed.

6. After unit has stopped from a cooling cycle or power interruption, the Yorkguard will not permit unit to start again for 5 minutes. (This protects compressor by providing refrigerant circuit equalization.)

7. If discharge pressure reaches 385 psig or discharge temperature reaches 275°F, the Yorkguard will stop outdoor unit and put system on lockout. The emergency light on thermostat will be energized, alerting homeowner that a malfunction has occurred.

8. To restore operation after lockout (if control causing lockout has automatically reset):

a. Turn thermostat dial beyond satisfied position and return to original setting, or
b. Turn thermostat system switch to "OFF" position and return to cooling position, or
c. Interrupt line power or 24-volt control circuit power to indoor unit. (This method will reset Yorkguard and subject unit to 5-minute delay.)

With Power to Unit and Thermostat in Heating Position

1. Reversing valve is deenergized to position refrigerant circuit for heating. Indoor coil functions as condenser and outdoor coil functions as evaporator.

2. If fan switch is in "ON" position, a circuit is made through blower relay providing continuous blower operation.

3. Solid-state Yorkguard starts time delay period. After 5 minutes, unit is ready for operation.

4. When first stage of thermostat heating contact closes, a circuit is made through the Yorkguard to energize contactor and start the unit. With fan switch in "AUTO" position, a circuit is made from thermostat heating contact through blower relay to provide blower operation.

5. Unit will cycle in response to thermostat signal to provide heating as needed.

6. After unit has stopped from a heating cycle or power interruption, the Yorkguard will not permit unit to start again for 5 minutes. (This protects compressor by providing refrigerant circuit equalization.)

7. If discharge pressure reaches 385 psig, discharge temperature reaches 275°F or system tries to go on defrost cycle, the Yorkguard will stop outdoor unit and put system on lockout. The emergency light on thermostat will be energized, alerting homeowner that a malfunction has occurred.

8. To restore operation after lockout (if control causing lockout has automatically reset):

a. Turn thermostat system switch to "OFF" position and return to heating position, or
b. Turn thermostat system switch to "OFF" position and return to heating position, or
c. Interrupt line power or 24-volt control circuit power to indoor unit. (This method will reset Yorkguard and subject unit to 5-minute delay.)

9. Supplemental electric heaters are energized by second stage of heating thermostat. Yorkguard will permit operation of supplemental heaters below balance point. At all outdoor temperatures above balance point, supplemental heaters are not permitted to operate.

10. Supplemental heater operation with outdoor temperature below balance point:

a. With second stage of heating thermostat contact closed, a circuit is made through Yorkguard to energize control sequencers. Control sequencers will energize supplemental heaters in 5-kW steps to permit power loading of electric lines in small increments. Each kW heater is protected by automatic reset thermal limit switch and a backup fusible link.
b. When second stage of heating thermostat becomes satisfied, contact will open to deenergize supplemental heaters. Control sequencers will deenergize supplemental heaters in 5-kW steps in reverse order of energizing.

Supplemental Heater Control

The module governs the operation of supplemental heaters according to the following settings and conditions:

1. According to the balance point determined as in the section entitled Balance Point Determination and Setting, page 35, the jumper wire on the solid-state control module is set at the balance temperature at this temperature or the next higher setting. This determines the outdoor temperature at which the supplemental heat can be energized. For lower operating cost, the supplemental heaters should not be energized if the heat pump can supply the required heat.
2. When using the Evap-Blower Indoor unit, the amount of supplemental heat needed at any outdoor temperature is the difference between the heat loss line and the heat pump capacity. In the example of the section mentioned above, at +10°F outdoor temperature a total of 30 MBH is required and the heat pump can supply 14.5 MBH. Therefore, 15.5 MBH must be supplied by supplemental heaters.

Besides the defrost cycle, the module's control of which is described in the next section, there are two special conditions for which the module forecloses compressor operation as follows:

Operation Below -10°F Outdoor Temperature

1. At -10°F outdoor temperature, compressor operation cannot be justified due to small amount of heat generated.
2. Yorkguard senses -10°F outdoor temperature and performs the following functions:
   a. deenergizes compressor circuit;
   b. energizes standby heat (if installed) under control of second stage of heating thermostat (supplemental heat remains under control of second stage of heating thermostat);
   c. operates indoor blower under control of first stage of thermostat.

Operation in Emergency Heat Position

When switch on the thermostat is placed in emergency heat position:

1. Emergency light is energized.
2. Compressor circuit is locked out.
3. Supplemental and standby (if installed) heaters will be controlled by first stage of heating thermostat.
4. Indoor blower will operate on demand for heat and cycle off with the last heater element when in "AUTO" position.

COIL FREEZING AND DEFROST

In the heating mode, the heat pump evaporator at certain conditions of ambient dry and wet bulb temperature experiences coil freezing. Data from General Electric show that the highest number of defrost cycles seem to occur at air temperatures between 32°F and 40°F. Coil freezing seems not to occur, on the other hand, at lower temperatures when the relative humidity is usually well below 60%. Coil freezing can be especially severe if sleet or rain blows onto the evaporator coil; however, little is known about conditions causing frost formation, rates of frost layer and density increase and its effect on heat transfer rates, although there has been some research on the subject (Gates et al., 1967; Huffman and Sepsey, 1967; Parish and Sepsy, 1972).

While a moderate amount of frost on the evaporator may enhance heat transfer through increased roughness-induced turbulence and increased surface area, too much frost or ice can block, or at least significantly reduce, the air flow over the coil, allowing suction gas temperatures to rise. Sensing this condition, a defrost cycle is initiated to remove the ice.

The most common method of defrosting the outdoor coil is by cycle reversal; that is, the reversing valve is actuated and the heat pump is switched from a heating to a cooling cycle, while resistance heating elements are energized to temper the cold air being input to the conditioned space. Defrosting is accomplished by the high temperature vapor from the discharge side of the compressor entering and condensing in the iced-over outdoor coil and requires, on the average, 3-5 minutes, depending on the method of defrost control, heat pump design and other factors. System controls are usually arranged to prevent prolonged or excessively repeated operation in the defrost mode. A variety of different schemes is used to control the defrost cycle (Shaffer, 1977). One control system uses a timer to initiate and terminate defrost operation at definite time intervals. A thermostat or pressure sensor in the outdoor coil provides a signal used to override the timer, that is, to terminate the defrost cycle early, as soon as the outdoor temperature has risen enough to indicate that the ice has been melted. The same elements are used to abort the start of the defrost cycle when the temperature outdoors (outdoor coil) is high enough so that defrost is not really needed. The timer keeps on running and simply makes the system take another look at the end of the next interval.

A second scheme for the initiation of defrost uses a control that senses the air pressure drop across the outdoor coil. The pressure drop rises as the coil frosts up to occlude the outdoor, fan-driven air flow and, at a preset point, the pump can be reversed. Defrost operations are terminated by a signal derived from outdoor refrigerant temperature (pressure). This method was used for awhile by Fedders with very poor results and has since been superceded. As described in Chapter 4, wind gusts and drifting snow could lead to both unnecessary initiation of the defrost cycle and its inoperability when needed.

A third method of control is the temperature differential between two thermostats, one located to measure the outdoor temperature, and the other placed in the evaporator coil. As frost accumulates, the temperature difference rises, and the derived signal is used to initiate defrost when it is needed and to terminate the operation when the temperature inside the outdoor coil rises to a predetermined value. The American Refrigeration Institute (ARI) has a standard test for defrost function. ARI standards specify automatic defrosting conditions as 35°F (dry bulb)/33°F (wet bulb). They further require that the system shall perform satisfactorily over a period of time with no more than 20% of the total time in the defrost mode.

All the reverse cycle defrosting schemes will reduce the heating capacity of the apparatus. The lowest (air) temperature for which the heat pumps are usually designed is 20°F, and heat output at this low temperature is commonly only half the rated 45°F heating value. The introduction of the defrost cycle further decreases the limited low-temperature capacity. Table 2-2 shows some typical deratings based on defrost operation at 20°F. In the defrost range, operation of the defrost cycle reduces the heat generally delivered by 10-25%. Of course, if the outdoor air were very dry, the defrost mode might not need to be activated.

Table 2-2. Derating of Heat Pump Performance due to Defrost Requirements (Shaffer, 1977)

Normal Capacity Rating 45° (out)/ 70° (in) and COP		Application Rating 20° (out)/70° (in) and COP		Reduction in Application Rating Due to Defrost	Capacity and COP 20°/70° with Allowance for Defrost	
(Btu)	(COP)	(Btu)	(COP)	(%)	(Btu)	(COP)
23,000	2.4	16,000	1.8	12.5	14,000	1.6
28,000	2.5	19,000	1.9	16.0	14,000	1.4
39,000	2.3	27,000	1.8	3.7	26,000	1.7
46,000	2.3	33,500	1.85	7.5	31,000	1.7
47,000	2.2	32,000	1.7	25.0	26,000	1.4
58,500	2.5	39,500	1.9	9.0	36,000	1.7
62,000	2.4	41,000	1.8	14.5	35,000	1.5

The York heat pump systems use the Yorkguard module to control the defrost cycle, initiating it for a combination of high-pressure differential and liquid temperature prior to suction at the outdoor coil, according to the following program:

1. A defrost cycle is initiated when defrost switch (air pressure differential) has been closed for 12 seconds. (This delay eliminates effect of wind gusts and liquid temperature is 39°F or lower.)
2. Yorkguard energizes defrost relay which:
   a. energizes reversing valve to switch refrigerant circuit to cooling cycle;
   b. stops outdoor fan; and
   c. energizes first step of supplemental heat to prevent cold drafts in conditioned space.
3. Defrost cycle is terminated when liquid line temperature exceeds 75°F or liquid line temperature exceeds 45°F for 5 minutes. (This permits defrost termination if wind velocity does not permit liquid line temperature reaching 75°F.)
4. Yorkguard deenergizes defrost relay to return unit to normal heating cycle.
5. If unit tries to defrost within 5 minutes after completing a defrost cycle, Yorkguard will stop unit and lock it out. See no. 8, page 41, for reset procedure.

Carrier initiates defrost only according to outdoor coil temperatures in their Chronotemp defrost system, which utilizes both time and temperature to determine when defrost will take place and for how long. Defrost does not take place at all when the temperature of the outdoor coil is above freezing. This would be a waste of electricity. Defrost occurs every 90 minutes only when the temperature of the outdoor coil is below freezing. The defrost cycle ends as soon as the outdoor coil temperature is raised to about 65°F. (This varies slightly in various Carrier models.) Tests have proved that when coil temperatures hit this level, defrost is always completed.

The defrost action occurs when both the timer and the outdoor coil thermostat call for it. The Chronotemp system then takes three steps:

1. It reverses the four-way valve, sending hot refrigerant gas directly from the compressor into the outdoor coil to melt off the frost.
2. It shuts off the outdoor fan to permit the coil to warm more rapidly. (Indoor fan continues to circulate air.)
3. It activates electric resistance heater or heaters in the indoor heat pump section to prevent cool drafts.

The thermostat attached to the outdoor coil is a two-stage type, which ends the defrost cycle when the coil temperature hits 65°F.

The defrost mechanisms that operate defrost for a set number of minutes can be wasteful of electricity and heating capacity. At about 40°F when ice accumulation is heaviest, defrost might take 6 minutes. However, at a temperature of 10°F, when only light hoarfrost is present, all ice might be off the coil in 3 minutes. Chronotemp automatically adjusts the length of the defrost cycle to these changing conditions. The fact that defrost ends as quickly as possible is especially important in very cold weather, when maximum heat output is most needed.

In very cold weather with a high wind, coil temperature might never go as high as 65°F. The Chronotemp timer takes care of this problem by automatically "overruling" the thermostat to end the defrost cycle after 10 minutes. Tests prove that defrost has always been accomplished at the end of this period. The heat pump is then prevented from starting up on defrost cycle. Thus, when the indoor thermostat calls for additional heat, it is always provided immediately, and defrost is deferred until the heat pump has been in operation for some time.

In Westinghouse systems, which use subcooling refrigerant control, the defrost cycle is initiated by a timer, subject to control by the temperature of the outdoor coil and terminated by the temperature of the refrigerant leaving the outdoor coil. At the end of each timing cycle, defrost is started only if the outdoor coil temperature is 26°F or lower. When, after defrost, the refrigerant leaving the outdoor coil reaches 80°F, the defrost switch moves to restart the outdoor fan and deenergize the reversing valve to place the refrigerant circuit back on heating.

Information on the defrost cycles was provided by company representatives as follows:

- York–Robert Watkins of York Distributors in Alexandria, VA
- Carrier–Steve Muhlheiser of the A/C Division in Syracuse, NY
- Westinghouse–Roche and Hull Inc., Distributors in Landover, MD

THE ROTARY COMPRESSOR

Although most of the unitary heat pump manufacturers utilize welded hermetic reciprocating motor compressors in their heat pump equipment, Fedders has introduced a split system product with a welded hermetic rotary motor compressor (Boxer, 1976). Advantages claimed for the rotary, as opposed to the reciprocating positive displacement compressor, are: (1) its simplicity of construction (fewer moving parts, more rugged rotor, single discharge valve, etc.); (2) its ability to better withstand liquid slugging; and (3) its denser suction gas, since suction gas motor cooling is not used. The greater amount of refrigeration oil claimed as a benefit for compressor protection can reduce refrigeration capacity and efficiency elsewhere in the system by decreasing evaporation rates and inhibiting heat transfer. Other disadvantages of the rotary compressor are: (1) the greater possibility of leakage of high-pressure gas to the low-pressure side past rotor and vane clearances; (2) greater heat losses to ambient air due to a "hot shell" design (in the case in question); and (3) the possibility of clearance volume increases (with attendant volumetric efficiency loss) due to vane and system wear. According to Kirschbaum and Veyo (1977), the rotary compressor is being considered by other heat pump manufacturers for possible adoption. The operational core of the Fedders compressor is pictured in Figure 2-10.

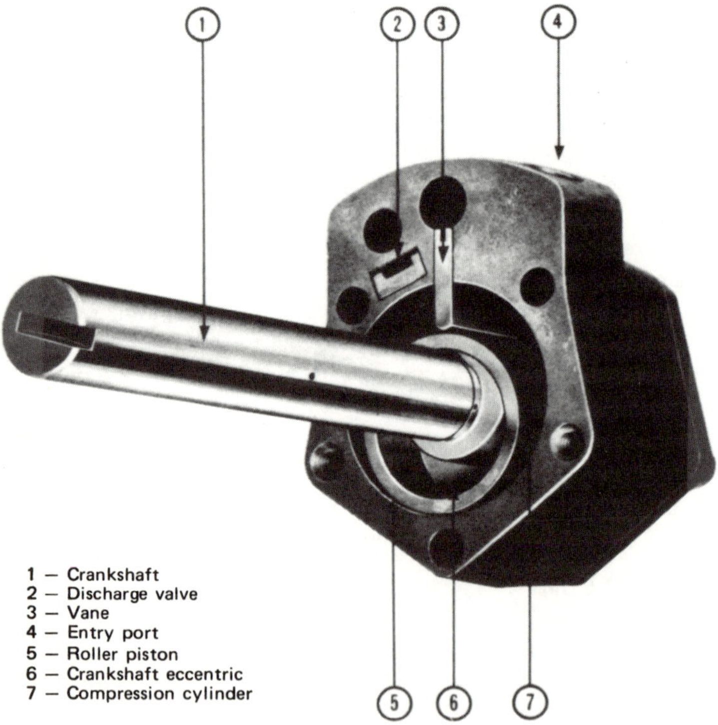

Figure 2-10. Core of Fedders rotary compressor (Fedders Compressor Company, Division of Fedders).

In the rotary compressor, the back-and-forth reciprocating motion of pistons in cylinders is replaced by the simple rotary motion of a crankshaft within a single compression cylinder. A "roller" piston and spring-loaded vane, which rests against it, divide the chamber into two volumes of continuously and inversely varying size.

On one side of the vane, from the tangent point of the roller piston to the cylinder wall, suction gas is taken into the chamber. On the other side of the vane, gas in the chamber is being compressed.

Thus, the internal geometry of the rotary compressor enables the suction and compression phases to proceed simultaneously and continuously. Almost 100% of the available displacement is utilized productively 100% of the time. The resultant high volumetric efficiency of the rotary compressor is reflected in its remarkably compact size and light weight. The rotary compressor weighs approximately half as much as a reciprocating compressor of comparable capacity and is from 40-50% smaller.

The principles of this operation are illustrated more concretely in Figure 2-11, which shows the operation of the inner eccentric valve

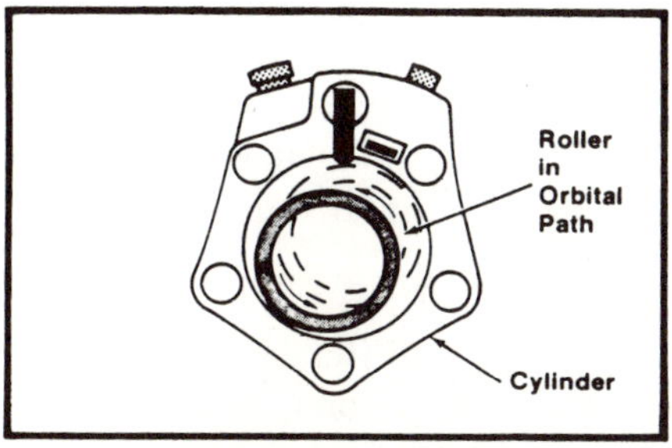

The cylinder receives the refrigerant gases and houses the compression components. Compression is accomplished by action of the roller rotating within the cylinder while traveling in an orbital path.

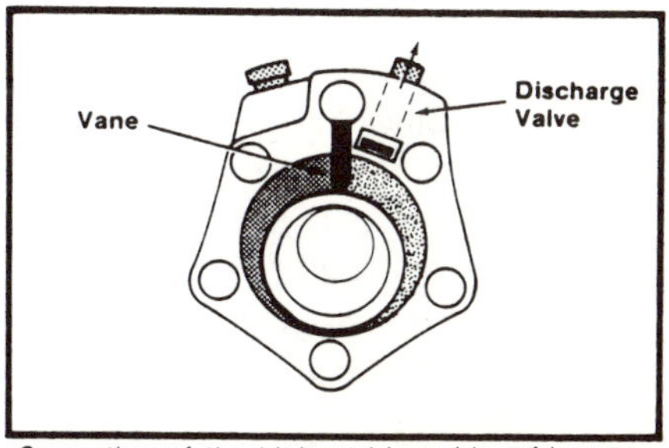

Separation of the high and low side refrigerant pressures, within the cylinder, is maintained by the vane and by an oil seal between cylinder components.

A discharge valve contains the high pressure gas within the cylinder until design cylinder pressure is reached.

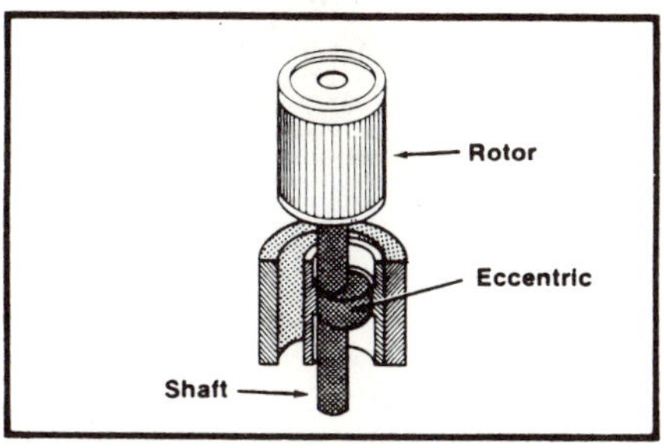

The roller is driven by an eccentric shaft which is shrunk to the rotor of the motor. This eccentric allows the roller to operate at a lower R.P.M. than the motor.

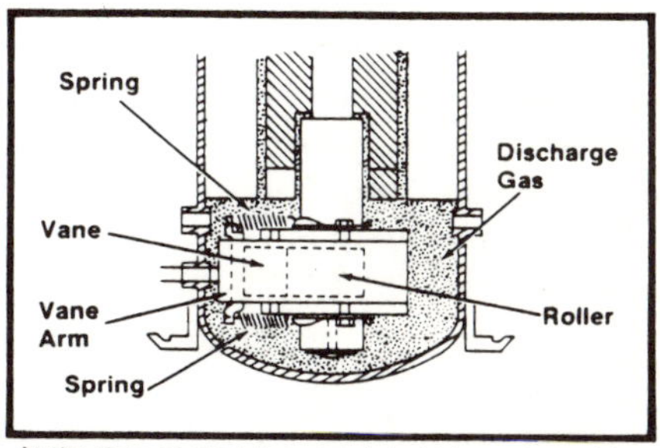

At the time of start the vane is held in contact with the roller by the action of the vane arm and two springs. Once the compressor is in operation, pressure of the discharge gas within the shell aids the springs in holding the vane in the proper position.

Figure 2-11. Principles of Fedders rotary compressor (Fedders Compressor Company, Division of Fedders).

vane and unique method of retaining the vane in contact with the eccentric roller.

A description of some of the details of producing and packaging the compressor at the Fedders compressor manufacturing plant in Frederick, Maryland, provides some insight into the problems of integrating design with production facility that must be overcome to successfully bring a new product into the market.

To begin with, the core must be made in a series of capacities and packaged in a clearly defined envelope so that Fedders can install the compressors in its line of heat pumps and air conditioners using standardized assembly techniques. Fedders uses four standard envelopes, termed frames, which increase in size with increments in both diameter and length at each step. Within each of these frames (Figure 2-12), the core cylinder is made to standard diameters for both the refrigerant chamber surrounding the shaft and the outer arc outside the vane. Variations in capacity within a frame are obtained by changing the depths of the cylinder. Since the core cylinder is precast as a simply connected region with the center hole represented by opposing depressions, there is just a little room for variation in their depth. Moreover, the drive motor would also vary in length.

The rough-cast core cylinder is machined first by grinding a pair of parallel flats defining the depth and reference planes. This leads to some variability in the outside arc, which is corrected by turning on a lathe, when necessary, to clear the outer surface inside the frame.

Induction heating is used to accomplish shrink fits in two instances. The stator for the drive motor is fixed with respect to the envelope by induction heating the envelope at the end where the stator is installed. Before the envelope cools, it is placed over the close-fitting stator. It then shrinks on cooling to grip the stator. A similar shrink-fit process is used to clamp the eccentric shaft into the rotor of the motor with precise linear spacing along the axis, so that it rotates at the proper position inside the core cylinder.

Practically all the machining and assembly operations are carried out by automatic machinery. The operator adjusts the machinery, then feeds the parts to be machined or assembled. Manual activities are most intense for the application of precision measuring equipment in the subassemblies, which require critical relative angular orientation among their constituent parts. Burrs on machined parts are removed by an agitated bath containing a slurry suspension of ceramic abrasive material. The envelope must be completely sealed by hand-soldered joints because it withstands the high discharge pressure of the unit. It is painted using electrolytic deposition from a water solution to ensure total penetration of all exterior crevices

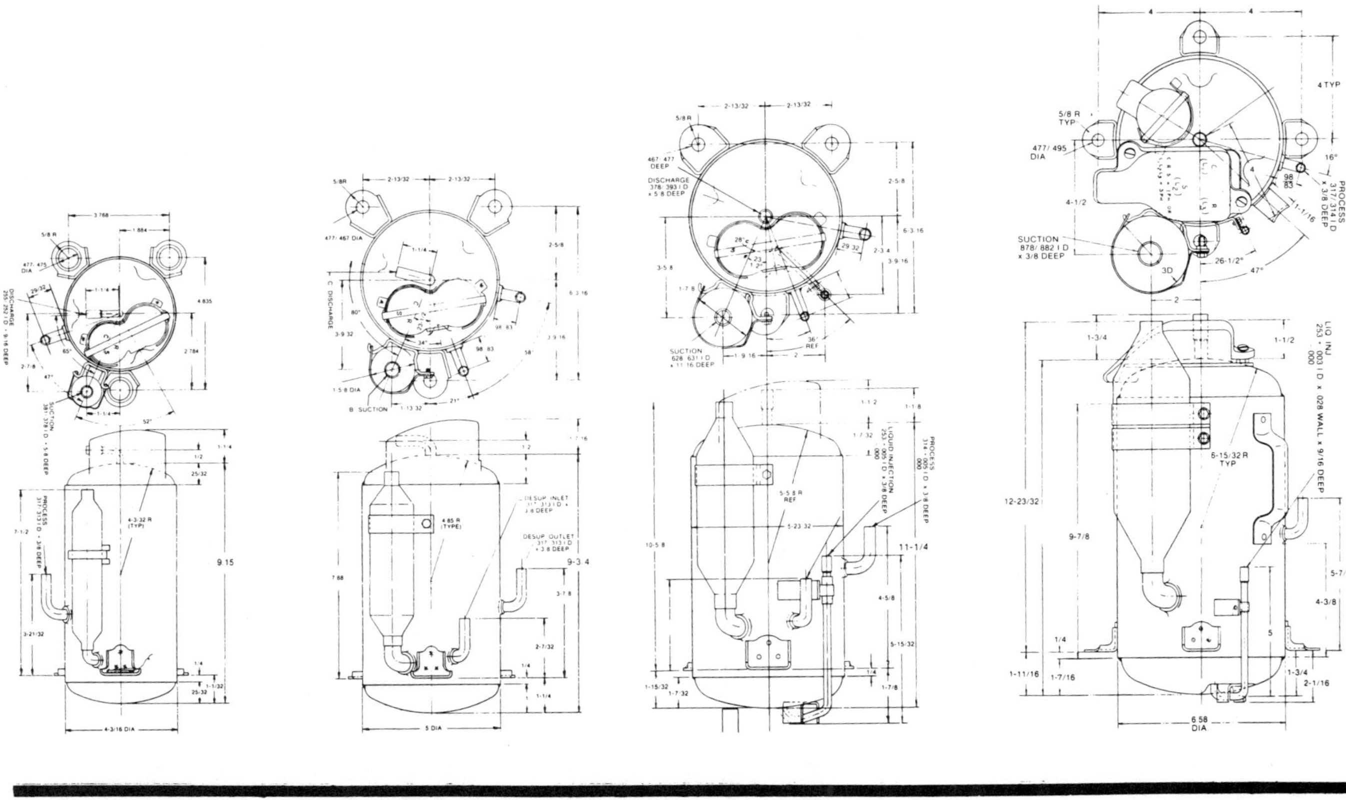

39 Frame 48 Frame 55 Frame 63 Frame

Figure 2-12. Envelopes for Fedders rotary compressors (Fedders Compressor Company, Division of Fedders Corporation).

to prevent rust initiation. Adhesion of print has been promoted by a special "bonderizing" process.

Because there are so few moving parts, tolerances needed for most interlocking parts are easily achieved in the machining process. There is one area of central selectivity–the vane that bears against the rotor–that must be very closely held in its slot. In this case, vanes are selected to match the slots in the cylinder block using precision gauges. Because the vanes are cheaper to produce, they are made to excess. In that way, there are enough vanes to be matched with the slots on the wings of their statistical distribution for the smaller number of cylinders.

Some of the advantages of the rotary compressors are as follows (Boxer, 1976):

There is no suction valve. Valve provisions consist simply of a discharge valve made of a U-shaped leaf of Swedish steel secured to a light steel backer; the entire assembly is fitted into a rectangular pocket in the cylinder block. Two such valves are used in compressors of higher capacity.

In operation, the compressor chamber is divided into two areas of constantly changing size and function by the vane, which rests against a film of oil on the roller. By contrast to the back and forth motion of a piston in a cylinder, the suction "stroke" in the rotary compressor changes imperceptibly into a discharge "stroke" at a midpoint in the cycle, with no alteration in speed or direction.

Meanwhile, on the opposite side of the vane, the discharge stroke (or more properly the discharge phase) is undergoing an equal, but opposite, change in size and function. Suction and discharge processes proceed continuously and simultaneously.

Because the suction phase is active at all times within the compression chamber, there is no need for a suction valve. The absence of a suction valve and the elimination of liquid slugging as a hazard provide special reliability at low ambient conditions.

Because the unbalanced components in the pump assembly–the roller and the eccentric on the shaft–move around the center of rotation, it is possible to balance them dynamically by counterweights on the lower end of the motor rotor, which is located above the pump assembly and concentric with it. This permits the motor to be rigidly mounted within the shell without any provision for internal spring suspension to isolate pump vibration from the rest of the compressor. The requirements for springing the pump assembly demand more space, a larger shell and more steel in a reciprocating compressor.

With the shell of rotary compressor shrink-fitted directly to it, the stator of the compressor motor becomes the limiting factor on size. Rotary compressors are thereby about one-third smaller in physical size than reciprocating compressors of comparable capacity.

In the rotary compressor, there are no internal baffles, accumulators or suction mufflers. Suction gas enters the compression chamber directly from an external suction line accumulator. By contrast to reciprocating compressors, in which suction gas is superheated as it passes over the motor before passing through the suction valve, the gas entering the compression chamber of the rotary compressor is denser and results in greater pumping rate per cubic inch of displacement. This permits a corresponding reduction in size of compression parts with its attendant advantages in lower cost, lower weight and less friction.

The passage of suction gas directly into the compression chamber and the discharge of high-pressure gas into the shell is a fundamental difference from the reciprocating compressor. In the conventional reciprocating compressor, the shell is on the low-pressure side of the system; the rotary compressor shell is on the high-pressure side. This arrangement permits the use of oil to seal moving compression parts without the tight clearances usually required. Oil is carried up into the compression chamber through the crankshaft in large volume and is separated out from the compressed gas when the mixture is discharged into the shell. In a reciprocating compressor, the drain on the crankcase oil supply, which would result when compressed gas is discharged into the condenser, makes this generous use of oil as compression seal impractical.

Because of effective use of oil as a seal, clearances in the rotary compressor have been considerably opened up, as have manufacturing tolerances. Selective fits have been practically eliminated.

A comparison of typical operating temperatures within a rotary and reciprocating compressor with a 45°F evaporating temperature and a 115°F condensing temperature, is shown in Table 2-3.

Table 2-3. Typical Operating Temperatures in Hermetic Motor-Compressors (Boxer, 1976)

Reciprocating Compressor	°F	°F	Rotary Compressor
Entering Suction Gas	45	45	Entering Suction Gas
Motor Heat	100	—	
	145		
Heat of Compression	70	70	Heat of Compression
Discharge to Condenser	215	115	Discharge to Shell
		100	Motor Heat
		215	Discharge to Condenser

The figures that present the greatest difference are the 45°F temperature of suction gas, which is used to cool the motor in the reciprocating compressor, and the 115°F temperature of the discharge gas, which cools the motor of the rotary compressor. Since the discharge gas in the rotary compressor is more than twice as dense as the suction in the reciprocating compressor, it has a much higher heat transfer coefficient to make up for the higher temperature.

The rotary compressor omits two troublesome heat pump features. First, it dispenses with off-cycle heat. With no suction valve in the rotary compressor, liquid slugging presents no hazard. Moreover, there were indications that liquid refrigerant in the crankcase, which was vaporized by off-cycle heat, could move directly into the condenser entraining oil along with it, since it did not have to pass through a discharge valve. Second, the rotary compressor does not provide a high-pressure relief. It is probably the only air conditioner in the 2- to 5-ton range listed by Underwriters' Laboratories (UL) without this protective device. Since the motor protector in a rotary compressor is located in the high-pressure side of the system, it reads and reacts directly to high pressure and temperature conditions. In a reciprocating compressor, the motor protector is unaware (except indirectly through rising temperatures) of dangerous pressures only an inch or two away on the other side of the discharge valve.

Where the motor protector does not shut the system down, the rotary compressor will unload itself in the compression chamber before pressures are one-third the way to the bursting point. The vane separating high- and low-pressure areas is simply forced back and the relief is achieved internally with no damage.

For application to include heat pumps, three carbon-graphite bearings were used to support the crankcase in place of sintered bearings, which had been sufficient for air conditioners. In the Fedders rotary compressor, carbon-graphite bearings are adequately lubricated by liquid and vaporized refrigerant in the event of temporary lubrication failures encountered on heating cycles. On special tests, a rotary-powered heat pump has operated for up to 12 days without any oil in the system at all.

Shortly after they were included in heat pump compressors, carbon-graphite bearings became standard equipment for all rotary compressors to reduce inventory and production costs.

OTHER SIGNIFICANT HEAT PUMP SYSTEM FEATURES

Installation and integration of the heat pump into existing air or water thermal media circulation systems requires attention to some significant features. One example of such installation is shown in Figure 2-13 for the

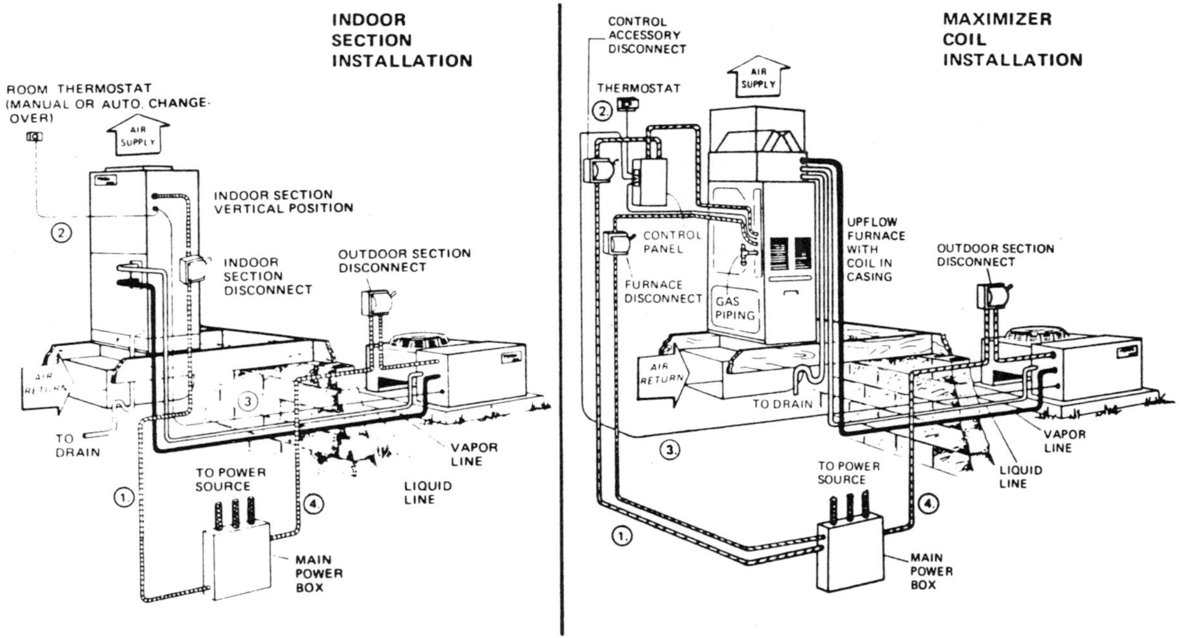

(1.) Power Wiring – Indoor Unit
Single Phase (– 6) – Two conductors + ground.
Three Phase (–25) – Three conductors + ground.

(2.)(3.) Low Voltage Wiring –
19 AWG 10-Wire Conductor Cable
York Part No. 472-00283A
(Maximum length = 75')*

(4.) Power Wiring – Outdoor Unit
Single Phase (– 6) – Two conductors + ground.
Three Phase (–25) – Three conductors + ground.

*Heavier gauge low voltage cables may be required for long lines.
For longer lengths refer to 1505.20-AD1.

Figure 2-13. Typical field wiring and piping for York heat pumps.

York heat pump. In the Maximizer, the heat pump is used instead of gas burning only when temperatures are most propitious for its COP. Gas burning serves as the supplemental heat in the formal heat pump sense, although in a severe winter it may provide most of the heat.

Generally, the location of the outdoor part of the heat pump will pose no problems for electrical wiring, which may be chosen solely in terms of its current-carrying capacity and voltage insulation requirements, as shown in Table 2-4 for the York heat pump. However, the tubing used to convey refrigerant from the compressor to the indoor heat exchanger must not introduce unduly high pressure loss or pose absolute limits on refrigerant flow capacity through its circuit. For example, York cites 75 feet (23.2 meters) as the maximum length of refrigerant tubing to be used under any circumstances. In addition, there are maximum differences in elevation between outdoor and indoor units allowed, depending on total line length according to Table 2-5. A related Table 2-6 by Fedders cites the differences in tubing diameter allowed according to total line length and capacity of the unit.

Pipe requirements for refrigerants are very similar for all U.S. heat pumps because they all use fluorocarbons, or some variety of freon, which have similar pressure density relationships affecting the flow properties through long tubes. Although gas cycles are possible, almost all actual air conditioning, refrigerating and heat pump circuits are based on the use of a condensable working fluid for the large heats of vaporization and condensation in the evaporator and condenser coils.

To maintain high rates of evaporation and condensation per unit of equipment, the systems must be designed for high mass transfer. To obtain this and to ensure efficient pumping at the same time, we would generally select a working fluid that has a relatively high vapor pressure at the design evaporator temperature. For a heat pump we might be most concerned with the range from 0–40°F. At the same time, high condensing pressures are to be avoided; otherwise, special high-pressure construction will have to be used. In practice, almost all packaged equipment is designed for a small positive pressure at all times, which eliminates the need for a purge system.

Common copper tubing is generally furnished in grades that have working pressures (including a generous safety factor) of 250–400 psi, and properly designed silver soldered assemblies made with such tubing will perform satisfactorily in this pressure range. Refrigerants that may be useful on the basis of suitable boiling points, modest pressures, reasonable performance, etc., are listed in Table 2-7, which has been divided into two sections: (A) gases of all kinds; and (B) fluorocarbons. These fluorocarbons have been discussed in detail by Hudlicky (1961). Table 2-8 presents their physical constants underlying the operational properties in Table 2-7. The gases from the

Table 2-4. Electrical Data—Outdoor Unit of York Heat Pumps (York Division of Borg-Warner)

Model		CHP024-6A	CHP030-6A	CHP036-6A	CHP042-6A	CHP048-6A	CHP060-6A	CHP042-25A	CHP048-25A	CHP060-25A
Unit Power Supply				208/230-1-60					208/230-3-60	
Voltage Min-Max						198-253[a]				
Max. Dual Element	230V	20	30	35		45	60	25		35
Fuse Size (Time Delay)	208V	25	30	35	40	45	60	25	30	40
Unit Ampacity	230V	17	22	26	29	33	52	18	20	28
	208V	18	24	27	31	35	52	19	21	30
Min. Field[b] Wire Size	230V	12		10		8	6	12		10
AWG 60°C, 3% Voltage Drop	208V	12	10		8		6	12	10	
Max. Wire Size AWG				8			6		10	

[a]Unit may operate satisfactorily to 187 volts if outdoor ambient is lower than 115°F.

[b]Wire sizes are good up to 75 feet in length. Use only copper conductors for connections at unit.

Table 2-5. Maximum Length and Lift of Interconnecting Tubing for York Heat Pumps (York Division of Borg-Warner)

Model	Factory Precharged Line Sizes Tubing (o.d. inches)		Maximum Elevation Difference					
			Indoor Unit Above[a]			Outdoor Unit Above		
			Total Length of Line (ft)					
	Liquid	Vapor	30 or Less	50	75	30 or Less	50	75[b]
CHP024		5/8	30	28	23	30	33	28
CHP030	5/16	3/4	30	25	18	30	31	25
CHP036			27	20	11	30	22	12
CHP042			30	28	23	26	22	17
CHP048	3/8	7/8	30	25	18	24	19	12
CHP060			27	20	11	20	14	8

[a]Maximum elevation of one unit section above the other based on continuous operation at 60°F outdoor air temperature. Continuous operation at lower temperatures may result in evaporator freezeup.

[b]Maximum length of tubing = 75 feet. Example: (15 ft + 30 ft + 30 ft) = 75 ft.

Table 2-6. Fedders Refrigerant Tubing Sizes[a] (Fedders Corporation)

Nominal Unit Tonnage	To 21 ft Suction (in.)	To 21 ft Liquid (in.)	22-39 ft Suction (in.)	22-39 ft Liquid (in.)	39-50 ft Suction (in.)	39-50 ft Liquid (in.)	50-75 ft Suction (in.)	50-75 ft Liquid (in.)
2	5/8	1/4	3/4	3/8	3/4	3/8	3/4	3/8
2½	5/8	1/4	3/4	3/8	3/4	3/8	3/4	3/8
3	3/4	3/8	7/8	3/8	7/8	3/8	7/8	3/8
3½	3/4	3/8	7/8	3/8	7/8	3/8	7/8	1/2
4	3/4	3/8	7/8	3/8	7/8	1/2	1-1/8	1/2
5	7/8	3/8	1-1/8	3/8	1-1/8	1/2	1-1/8	5/8

[a]Boxed area represents lengths or sizes in excess of available accessory tubing kits for converting 3/4 in. to 7/8 in. and 3/4 in. to 1-1/8 in. tubing. For tubing lengths in excess of 50 ft, add 3 oz of oil for every additional 10 ft. Refrigerant charge adjustment for tubing lengths beyond 25 ft: (a) 3/8 in. o.d. liquid line–add 0.6 oz F22 for each foot beyond 25 ft. (b) 1/2 in. o.d. liquid line–add 0.6 oz F22 for each foot up to 25 ft and 1.2 oz for each foot beyond 25 ft. (c) 5/8 in. o.d. liquid line–add 1.2 oz F22 for each foot up to 25 ft and 1.8 oz for each foot beyond 25 ft.

general list are much cheaper than the fluorocarbons and are preferred in some industrial/commercial applications. Those on list A are all noxious, poisonous, flammable and explosion hazards, so are not generally considered safe for home appliances or for direct contact with the air circuits in buildings that may contain any significant number of people. The dangers of using ammonia may actually have been overestimated. It is poisonous, to be sure, but not acutely so; and, according to the U.S. Department of Commerce Census of Manufacturers, in 200 million hours of operation logged before 1970, only four fires or explosions were associated with its use.

The fluorocarbons are by far the most widely used refrigerants. Current U.S. fluorocarbon sales run $435 million; about half the U.S. production goes into propellant formulations and this usage may be curtailed due to its possible adverse affect on the normal chemistry of the upper atmosphere. There is also another possible toxic effect, which has been pointed out by Patton (1979), describing a campaign being carried out by Dr. H. Normal Runsdorf of New York who believes refrigerants exposed to high temperatures sparked some outbreaks of so-called Legionnaire's disease in 1976. The physical basis for the suspicion is thermal dissociation of fluorocarbons into phosgene when making contact with a flame, as originally reported by Midgely and Henne (1932). It is expected that allowed fluorocarbons would

Table 2-7. Characteristics of Commonly Available Gases and Fluorocarbons Suitable as Refrigerants (Shaffer, 1977)

	Evaporator Pressure (psig)	Condensing Pressure (psig)	Comp. Displacement cfm/ton	Horsepower/ ton	COP Rankine Cycles
A. Commonly Available Gases					
Propane	27.2	140.5	4.09	1.03	4.6
Ammonia	19.6	154.5	3.44	0.99	4.8
Methyl Chloride	6.5	80.0	5.95	0.96	4.9
Isobutane	3.3 (vac. in	44.8	11.5	1.08	4.4
Sulfur Dioxide	5.9 mercury)	51.8	9.1	0.97	4.9
Methylamine	9.9	46.8	10.2	0.98	4.8
Butane	13.2	26.9	15.5	0.95	5.0
Ethyl Chloride	20.5	12.4	24.8	0.91	5.2
Ethyl Amine	23.1	10.0	38.7	0.85_5	5.5
Tabulation based on 5°F evaporation/86°F condensation					
B. Fluorocarbons					
R-502 R-22/R-115 Azeotrope	36.0	175.1	3.61	1.08	4.4
R-22 Chlorodifluormethane	28.2	158.2	3.55	1.01	4.7
R-115 Chloropentufluoroethane	24.0	135.8	5.30	1.17	4.0
R-500 R-12/R-152a Azeotrope	16.4	112.9	4.95	1.01	4.7 (4.6_5)
R-12 Dichlorodifluoromethane	11.8 (vac. in mercury)	9.33	5.83	1.00	4.7
R-114 Dichlorotetrafluoroethane	16.1	22.0	20.1	1.05	4.5
R-21 Dichlorofluoromethane	19.2	16.5	20.4	0.94	5.0
R-11 Trichlorofluoromethane	23.9	3.5	36.5	0.94	5.0

Tabulation based on 5°F evaporation/86°F condensation.

be used in systems, which, although they may leak occasionally, are intended to contain the product. About 28% of the U.S. fluorocarbon production appears as the refrigerant in residential and commercial air conditioners, refrigerators, food storage and display, and auto air conditioners. The most widely used of the fluorocarbon refrigerants are: R-11, R-12 and R-22. These three account for nearly 92% of the U.S. refrigerants consumption.

Table 2-8. Physical Properties of the Freon Fluorocarbons [DuPont Catalogs, FM-1 (1965), BE-2 (1964)]

Freon	R-11	R-12	R-21	R-22	R-114	R-115
Formula	CCl_3F	CCl_2F_2	$CHCl_2F$	$CHClF_2$	$(CClF_2)_2$	$CClF_2$-CF_3
Molecular Weight	137.38	120.93	102.93	86.48	170.94	154.48
Melting Point, °C	-111	-158	-135	-160	-94	-106
Boiling Point, °C	23.8	-29.8	8.9	-40.8	3.8	-38.7
Liquid Density, g/ml at °C	1.476/25	1.311/25	1.366/25	1.194/25	1.456/25	1.291/25
Heat Capacity, Cal/g °C:liquid at 25°C	0.209	0.240	0.256	0.300	0.243	0.285
Vapor at Constant Pressure, 1 atm at 25°C	0.135	0.145	0.140	0.157	0.170	0.164
Latent Heat of Vapor, at bp Cal/g	43.51	39.47	57.86	55.81	32.51	30.11

As a guide to general practice regarding installation and testing of heat pump systems, the American Refrigeration Institute arrived at some general characterizations, which are followed in the service manuals of leading heat pump manufacturers. Standard ratings, operating conditions and performance tests for heat pumps generally follow the outlines of Tables 2-9 and 2-10. Suggested tools for installation, testing and service are listed in Table 2-11.

Typical costs for a heat pump system, fully installed, will, of course, depend on the size of the unit, existing internal system of the residence, and negotiations with the dealer. For units of 2- to 5-ton capacity (24,000-60,000 Btu/hr or 6,000-15,000 kcal/hr), the fully installed prices will probably range from $2,000-$10,000 in 1978. Some idea of representative prices for parts and price trends is shown in Table 2-12.

A fairly comprehensive study of comparative installation and maintenance costs was carried out by Gordian Associates (1978) for the 1976-77 period. Installation costs, with their variation by region, are shown in Table 2-13 averaged over all heat pump installations in 1977. Since the residential heat pump capacity distribution is closely clustered about 3 tons (~ 9000 kcal/hr), these costs pertain to that capacity. They include, in addition to the heat pump system, the insulated and noninsulated ductwork for gas and oil systems and, as required, the cost of a chimney and oil tank. Aggregating these costs along with the air conditioner tends to obscure the stark contrast between gas/warm air and the heat pump

Table 2-9. Standard Ratings, Heat Pump Systems (American Refrigeration Institute)

Standard Capacity Rating Definitions

A Standard Rating is a rating based on tests performed at Standard Rating Conditions. The test methods and procedures are described in thc ARI Standard.

Performance Tests

Five[a] performance tests must be passed. The indoor-air quantity must correspond to the Standard Rating maintained throughout the Maximum Operating Conditions Tests and at the beginning of the other three tests.

Maximum Operating Conditions Test. (Heating and Cooling) Tests are run at 90%[b] and 110% of rated voltage. The unit is operated continuously for 2 hours at the temperature conditions and voltage(s) specified; power is cut off for a period sufficient to cause the compressor to stop (not to exceed 5 sec), and then restored. Requirements are:

1. Operation must take place without damage to any part of the unit;
2. There must be continuous motor operation for the first 2 hours without interruption;
3. The heat pump shall resume continuous operation within 2 hours of restoration of power and shall then operate continuously for 1 hour. Operation of safety devices prior to the establishment of continuous operation is permitted.
4. Satisfaction of requirements without exceeding 15 psi water pressure drop through the unit.

Low-Temperature Operation. (Cooling) The unit shall operate continuously or stop and start under automatic limit device for 12 hours. Requirements are:

1. Operation must take place throughout without damage.
2. Frost or ice formation must not be sufficient to cut off air flow.
3. During test and subsequent defrosting, all ice or meltage must be caught and removed by the drain.

Automatic Defrosting Test. Air-source units, designed for operation in the frosting range, shall operate continuously for at least 3 hours. Requirements are:

1. Defrosting time shall not exceed 20% of total test time.
2. Indoor supply air temperature from unit shall not drop below 65°F for more than 1 minute.

Application Ratings

Many applications of unitary heat pumps require ratings under conditions other than conditions specified above.

Specification sheets may show such application ratings, provided they include, or are accompanied by, the Standard Ratings plus the Required Application Rating, Heating, clearly designated as such, including a statement of the conditions at which the ratings apply. When the Required Application Rating, Heating is at an outdoor temperature above 20°F, it shall be designated as the lowest recommended operating outdoor temperature.

[a]Four for models other than air-source units.
[b]Except for certain 208-volt machines.

Table 2-10. Operating Conditions for Standard Rating and Performance Tests

Test	Indoor Coil Air Entering		Outdoor Surface: Water Source		Outdoor Surface: Air-Source Entering	
	Dry Bulb °F	Wet Bulb °F	In °F	Out °F	Dry Bulb °F	Wet Bulb °F
Standard Rating Conditions, Cooling	80	67	75	95	95[a]	–
Standard Rating Conditions, Heating[b]	70	–	60	–	45	43
Required Application Rating Conditions, Heating (air-source models only)	70	–	–	–	20[c]	19
Maximum Operating Conditions, Cooling Cycle[d]	95	71	80	100	115[a]	–
Maximum Operating Conditions, Heating Cycle	80	–	75	–	75	65
Low-Temperature Operation, Cooling	67	57	–	70	67[e]	–
Insulation Efficiency	80	75	–	80	80	–
Automatic Defrosting Conditions (air-source models only)	70	60 (max)	–	–	35[f]	33 (min)

[a]75 wet bulb (wb) maintained if condensate rejected to outdoor coil air.

[b]Water flowrate (gpm) as determined in Standard Rating, Cooling Test.

[c]Or lowest operating temperature recommended by manufacturer, if above 20°F dry bulb (db) (and so indicated by the manufacturer in his specifications and literature) and at 85% relative humidity.

[d]Surrounding air temperature may optionally be 115°F or 95°F db.

[e]57 wb maintained if condensate rejected to outdoor coil air stream.

[f]Or lowest operating temperature recommended by manufacturer (if above 35°F) and so indicated by the manufacturer in his specifications and literature, and at 80% relative humidity (minimum).

Table 2-11. Tools and Instruments Required for Service of Heat Pump Systems[a]

Minimum per Serviceman

Service Manual
Spirit Level (Suggested: 3 in. pocket size)
Sling Psychrometer
Three Pocket Thermometers (Suggested: two -30–120°F, one 0–220°F)
Flaring Tool
Tubing Cutter
Swedging Tool Set
Pinch-off Tool
Refrigeration Service Valve Set (Hermetic Service Kit)[b]
Pressure Gauge Manifold (with two shutoff valves)
Low-Pressure Gauge (suggested: 30 in. vacuum to 80 psi & 250 psi retard)
High-Pressure Gauge (suggested: 0–400 psi)
Three Charging Lines
Two Access Valve Core Remover & Replacer Tools[b]
Two Schrader-Type Valve Adapters (Robinair No. 14217 or equal)[b]
Service Valve Wrenches (quantity as necessary, rachet type preferred)[b]
Nut Driver Set
Miscellaneous Screw Drivers
Electric Solder Gun
Electric Hand Drill
Miscellaneous Hand Tools
Oil Can (with flexible spout)
Terminal Connector Kit (for attaching wiring terminals)
Refrigerant Leak Detector (electronic or halide torch, electronic preferred)
Volt Meter (range for heat pump)
Amp Meter (range for heat pump)
Ohm Meter (range for reading compressor motor winding resistance)
Fuse Puller

Minimum per Servicing Firm

Oxygen-Acetylene Brazing Outfit (suggested: compact type)
Dry Nitrogen: Cylinder (suggested: oil-pumped grade); Pressure Gauge; Regulator
Refrigerant Charging Cylinder (calibrated in ounces or fractions thereof for applicable refrigerants)
Portable Draft Gauge (type, range and calibration)[c]
Tools and Equipment for Cleanup after Hermetic Motor Burnout for Filter-Drier Methods:[c]

1. Permanent in suction line, or
2. Temporary in suction line

Acid Test Kit–Access to an Acid Test Kit is Required[c]
Renewal Parts Manual
Special Tools/Instruments/Manuals[b]
High-Vacuum Pump (suggested: 2 cfm capacity or more; 5-μ or less drawdown)
Thermocouple Vacuum Gauge (calibrated in microns)
As Necessary or recommended for Heat Pump Manufacturer's Equipment[d]

[a]As necessary or recommended for heat pump manufacturer's equipment.
[b]Special tools/instruments/manuals.
[c]Four for models other than air-source units.
[d]Except for certain 208-volt machines.

Table 2-12. Price Trends for Important Components and Subassemblies[a]
(*Air Cond. Heat. Refrig. News,* 17 March 1975)

Item	1975 ($)	1974 ($)	Middle 1973 ($)
Compressor (Mfr. A 2-hp single phase) per unit	135.61	117.00	107.65
Compressor (Mfr. A 3-hp single phase) per unit	162.96	143.10	132.38
Compressor (Mfr. B 2-hp single phase) per unit	351.80	355.41	336.54
Compressor (Mfr. B 3-hp single phase) per unit	440.00	425.00	403.30
"Freon" R-22-25 lb Cylinder	35.13	26.39	23.25
Freon R-12-30 lb Cylinder	30.34	21.72	18.37
Copper Tubing (0.25 in.) 50-ft roll	7.64	8.94	6.23
Copper Tubing (0.625 in.) 50-ft roll	20.52	24.49	17.22
Copper Tubing (1.125 in.) 50-ft roll	53.42	64.36	46.50
Silver Solder–1 lb (15% silver)	19.51	22.35	11.12
Silver Solder–1 lb (5% silver)	9.48	9.82	5.27
Wire (No. 18-3 thermostat) 500 ft	19.23	16.63	10.12
Wire (No. 18-5 thermostat) 500 ft	31.49	27.89	19.90
Filter (20 x 20 x 1 in.)	0.52	0.53	0.59
Filter (20 x 20 x 1 in.)	0.58	0.58	0.64
PVC (schedule 40.75 in. tubing) 1 ft	0.175	0.175	–
Galvanized Sheet (24 gal)/lb	0.26	0.25	0.135
Galvanized Sheet (14 gal)/lb	0.23	0.225	0.125
Stainless Steel (26 gal)/lb	1.11	1.13	0.685
Stainless Steel (14 gal)/lb	0.825	0.8375	0.52
Galvanized Coil (24 gal)/lb	0.2575	0.24	0.125
Galvanized Coil (14 gal)/lb	0.23	0.2125	0.1225
Copper Sheet/lb	1.30	1.83	0.95
Aluminum Sheet (0.040, 0.050, 0.063)/lb	0.64-0.72	0.56-0.64	0.35-0.41
PVC (0.75-in. sheet) ft	–	–	0.8
Duct Liner (glass fiber, 4 ft x 100 ft x 1 in. roll)	47.97	45.55	39.75
Duct Board (glass fiber)/ft	0.28	0.235	0.21
Duct Wrap (glass fiber, 4 ft x 100 ft x 1 in. roll)	32.94	24.80	26.80
Spiral Duct (6 in.)/ft	1.01	0.97	0.74
Spiral Duct (10 in.)/ft	1.58	1.45	1.19

[a]Trade prices were gathered in a survey of Florida wholesalers by *Florida Forum*, a publication of the Florida Roofing, Sheet Metal and Air Conditioning Contractors Association.

Table 2-13. Conventional Electric Heat Pumps and Other Systems: Comparison of Installed Costs, First Quarter, 1977 Dollars (Gordian Associates, 1978)

	Houston	Birmingham	Atlanta	Tulsa	Philadelphia	Seattle	Columbus	Cleveland	Concord
Electric Heat Pump–Standard	3255.24	3119.14	3079.57	3397.40	3304.03	2671.29	3234.22	3406.22	2749.34
Hybrid Electric Heat Pump–Gas Furnace	3645.17	3491.91	3453.25	4051.71	3719.87	3260.69	3628.84	3841.92	3200.87
Hybrid Electric Heat Pump–Oil Furnace	4489.03	4367.51	4292.81	5069.55	4564.95	4034.35	4468.40	4685.62	4039.59
Gas Warm Air Furnace/Central Air Conditioner–High Efficiency	2745.00	2602.43	2552.58	2947.48	2668.18	2345.43	2717.19	2881.25	2365.95
Oil Warm Air Furnace/Central Air Conditioner-High Efficiency	3614.87	3560.62	3437.51	3832.08	3545.02	3148.33	3580.08	3754.58	3211.11
Central Electric Furnace/Central Air Conditioner–High Efficiency	2882.93	2746.14	2717.75	3090.18	2850.39	2429.62	2899.34	3060.90	2508.79
Baseboard Convectors/Room Air Conditioners	2378.14	2264.37	2165.37	2165.29	2283.81	2452.54	2367.93	2562.84	2305.43
System with Uninsulated Ducts[a]									
Gas Warm Air Furnace/Central Air Conditioner–High Efficiency	2558.83	2546.26	2496.41	2981.82	2612.01	2234.26	2634.26	2825.08	2282.18
Oil Warm Air Furnace/Central Air Conditioner–High Efficiency	3458.70	3504.45	3381.34	3803.42	3488.85	3037.16	3496.93	3698.41	3127.34

[a]Current practice.

Table 2-14. Annual HVAC System Maintenance Costs
[(National Environmental System Contractors Association (NESCA) Mid-1976 Prices)]

System	Residential Systems Annual Maintenance Cost ($)
Heat Pump	149.60
Gas Furnace and Central Air Conditioner	104.50
Electric Furnace and Central Air Conditioner	109.70
Oil Furnace and Central Air Conditioner	127.90
Electric Baseboard Resistance Heaters and Room Air Conditioner Units	110.00
Room Air Conditioner[a]	21.00

[a]Sears-Roebuck Co. average price per air conditioner unit.

less air conditioner difference left plain in the comparison by Pickelsimer (1977). Cost of room air conditioners, which can be considerably cheaper and more efficient than a central air system, was based on the Sears Roebuck 1977 catalog. Maintenance costs averaged over all heat pump installations are shown in Table 2-14. They are meant to represent the average retail price of a full service contract to the consumer. Maintenance costs did not vary enough with respect to region to warrant considering the difference in overall cost comparisons. All the costs in these tables have probably escalated by about 7-8%/yr since 1977.

CHAPTER 3

TRENDS IN HEAT PUMP TECHNOLOGY

INTRODUCTION

While overall increases in efficiency are generally valuable, the improvement of low-temperature capacity as well as the COP of the heat pump is essential for more widespread applications in colder regions.

The sources of thermodynamic inefficiency in a Rankine cycle heat pump refrigeration loop have been well-documented (Ambrose, 1974), as have the general technical approaches to overcoming them. Recommended approaches to increased efficiency are as follows:

1. Increase efficiency of refrigeration loop components, decrease heat leaks and fluid pressure losses, e.g., improve compressor and fan motor efficiency, increase heat exchanger effectiveness, etc.
2. Improve thermodynamic "quality" of heat sources, e.g., evaporator-side thermal storage, air conditioner condenser evaporative precooling, etc.
3. Improve capacity modulation of heat pump to better conform to building heating and cooling load.

Specific component or other improvements being proposed usually fall into one of the above categories.

In most cases, the technology for the above improvements is either available or could be readily developed provided there was sufficient support for the research and development and that the higher initial cost of the more efficient equipment would not be a deterrent to purchase. In some cases, improvements must await the development of some critical device or system, e.g., a variable-speed mechanical gear system for a hermetic reciprocating compressor.

A rather comprehensive analytical study of electrically driven heat pump efficiency and reliability improvement was completed by Westinghouse under EPRI sponsorship (Kirschbaum and Veyo, 1977). The study details the possible "hardware" changes required for an optimal (in a life cycle cost sense) electric heat pump for Albany, New York/Boston, Massachusetts/ Denver, Colorado, climates (although some possible improvements are not described due to "patent interest"). Most of the recommendations center on the major components of the heat pump system as outlined immediately below. In addition, there are new ways of system organization centered on the source thermal reservoir that promise specific relief for the low ambient temperature problem. The most recent and current utility experimental programs, as outlined in Chapter 5, tend to emphasize this approach. Some of the precedents are described below, as are some experiments in the design and use of gas-driven heat pumps. The aim is not to improve heat pump mechanical working efficiency, which is barely affected, but source energy conservation through bypassing the losses attendant in central utility electricity generation and distribution, and winter utilization of the engine exhaust heat. Early developments in using chemically driven thermal absorption process for heat pumps are also described below, as far as possible, as much of this technology is still held very closely.

COMPONENT DESIGN

The major components identified for substantial gains in working heat pump cycle efficiency are the compressor, heat exchanger and outdoor fan.

Compressor Modifications

Hermetic compressor development is directed toward: (1) improved reliability, (2) improved COP at all load conditions, and (3) improved low-temperature heating capacity. The paragraphs below summarize inherent difficulties and the most promising concepts that have been advanced.

Hermetic compressor losses, either mechanical or electrical in nature, are:

- valve, valve port, suction and discharge tube and muffler pressure drops;
- internal leakage of discharge gas to suction side of compressor;
- friction, hermetic motor and compression cylinder conduction heat losses to suction gas;
- valve inefficiencies (mechanical);
- oil circulation work loss;
- clearance volume (compressed gas reexpansion loss); and
- anisentropic compression.

Mechanical friction losses and other inefficiencies can be reduced by valve and valve port redesign, reduction of bearing forces on contact surfaces, etc. It is doubtful there is much room for improvement in the above areas, given a competent design. In addition, better matching of motor torque-efficiency characteristics to loads encountered, especially in colder-climate heating operation, use of heat shields between the compressor high- and low-pressure sides, and use of ringed instead of plug-type pistons are already being implemented or considered (Kirschbaum and Veyo, 1977). Minor decreases in clearance volume may be achieved by cylinder and piston redesign; however, it is doubtful that much can be done to improve isentropic efficiency, i.e., a closer approach to a reversible adiabatic process, given the basic reciprocating compressor configuration and materials of construction, since strength, durability, ease of fabrication and cost are also important considerations.

Improved hermetic motor efficiency can also improve system COP and capacity by reducing heat leakage to suction gas and the consequent loss of volumetric efficiency. Motor maximum efficiencies are usually 85% or less for single-phase and 90% or less for three-phase motors normally used in heat pump compressors. Motor efficiency can be improved by increasing the "stack" height, providing more capacitance, using a better (higher magnetic permeability) grade of steel in the stator and by a number of other modifications (Johnson, 1974). Kirschbaum and Veyo (1977) discuss the efficiency-torque characteristics of single- and three-phase hermetic motors. They imply that it might be possible to get a closer match between the torque at which motor maximum efficiency occurs and the load conditions representing most of the heating cycle operating hours of the heat pump. However, it appears that little opportunity exists for substantial electric motor efficiency improvement without undue additional first costs. If, in addition, cost considerations force manufacturers to substitute aluminum wire for copper in stator windings, lower motor efficiency would result because of a higher resistive loss for the same field density. Three-phase motors have higher efficiency than single-phase motors; however, their use in equipment for residential applications is impractical in most cases.

One possible method for motor efficiency improvement is use of ceramic-metallic composite core materials of higher magnetic permeability than the iron in use today. However, discussions with one compressor manufacturer, Fedders, indicates that this technology is at least five years away.

Capacity Modulation of Compressors

Only limited further efficiency improvement appears possible with the basic constant-speed hermetic reciprocating compressor design. Nevertheless, even if these improvements were realized, there remains the basic problem of the mismatch between the variation with ambient temperature of typical building heat losses (or gains) and the heating (or cooling) capacity of the vapor compression cycle heat pump. This mismatch is the cause of two basic types of efficiency losses: (1) that due to the necessity for supplementing the compressor at low outdoor temperatures with resistance heat at a COP of 1; and (2) that due to equipment cycling* under part-load operation. The former has long been recognized as accounting for energy consumption of buildings heated with heat pumps. Evidence of the latter has only recently been published as a result of measurements being carried out at the National Bureau of Standards (NBS) (Kelly and Bean, 1976; Parker et al., 1977) and is not ordinarily factored into energy estimation calculations. Hence, one method of heat pump efficiency improvement is capacity modulation to conform with actual heating and cooling load variation with ambient temperature.

One method of compressor capacity modulation is by speed-halving, i.e., by switching from a two- to a four-pole motor configuration. In 1973, Lennox Industries introduced into their larger tonnage, unitary air conditioner line (*Lennox News*, 1973) a two-speed hermetic compressor, which generates 15 tons of refrigeration at a synchronous speed of 3600 rpm and 7.5 tons at a synchronous speed of 1800 rpm.** Claimed advantages for this method of capacity control include reduction in mechanical friction and refrigerant gas pressure drop losses, producing more than a 50% decrease in power requirements as capacity is halved (Cawley and Pfarrer, 1974). An additional benefit appears to be increased life of refrigeration oil due to lower average discharge temperatures in two-speed mode operation. The Lennox units are currently being tested by Texas Power and Light Co. for use with a solar collector for auxiliary heat (see Chapter 5). The Carrier Corporation is already marketing a two-speed system as their 38TQ series. There appears to be some divergence in results between the two designs. As reported by Texas Power and Light, the Lennox high-speed operation is slightly lower in COP than the low-speed operation, at both 17°F and 47°F. The Carrier unit behaves in the opposite manner. However, the Carrier system control logic is to use low speed only for temperatures above 35°F and high speed for lower temperatures, which achieves a better match with respect to refrigerant pressure differentials across the compressor.

---

*Not all of the transient character of the capacity can be attributed to the compressor, of course. Heat exchangers, fans, thermostats all have their transients as well. Another cause of efficiency loss is the effect of frost on heat transfer.

**Rated at 45°F evaporator, 125°F condenser, 65°F return gas and zero subcooling temperature.

Other methods of capacity modulation have also been employed in commercial-size unitary air conditioners. These include cylinder unloading by suction valve lifting mechanisms, discharge gas bypass to the evaporator and compressor discharge-to-suction side pressure equalization. The second method, it should be noted, bypasses the condenser and produces no power reduction at all (Cheney, 1974). While these methods may reduce the compressor pressure ratio and, thus, the work of compression, only speed variation of the methods discussed can reduce mechanical and fluid friction. Thus, the two-speed compressor is reported to improve compressor COP by 3.3%, whereas a cylinder unloaded compressor sustained a 32% COP loss in going from full- to half-capacity operation (*Lennox News* 1973). Exxon is developing for full-scale marketing a method of controlling speed in synchronous motors without power loss (*New York Times*, 19 May 1979; *Wall Street Journal*, 21 May 1979), based on an efficient alternating current synthesizer invented by Richard H. Baker. Such a method of compressor control would be far simpler and more effective than all of the foregoing and would all but eliminate the need for on-off cycling, thereby contributing a substantial gain in cold-climate seasonal COP.

Rotary Compressors

The advantages of using rotationally operating machinery rather than engines that use linear reciprocating motion are always attractive. The Fedders unit described in Chapter 2 is clearly an example. However, its operation depends on the linear reciprocating motion of the small vane bearing against the eccentric roller. To operate properly, this vane must maintain its physical integrity to a high degree. It is selectively fitted in the slot initially and especially hardened so that its surfaces will not wear. The vane must always ride in the slot perpendicular to the eccentric-bearing surface. Any undue friction at this contact point will exert a bending couple on the vane. Any yield to this will then cause the vane to bind in the slot and abruptly halt the rotary cam motion. This kind of risk motivates exploration of alternative rotary compressor principles in directions that may well incur new risks. For example, work is progressing on very small ultra high-speed turbine compressors. Small turbine compressors, perhaps only 2-3 inches in diameter, will run at speeds as high as 70,000 rpm. A special solid-state inverter will be needed to provide variable frequency high-speed drive, and it is thought that superior efficiency can be achieved by using a second small turbine for the expansion step. Coupling could be either electrical or mechanical (Shaffer, 1977). Once developed on a laboratory prototype, the question will then be how to achieve production efficiency of the kind already demonstrated in the Fedders program.

Heat Exchanger Effectiveness

Since thermal effectiveness is a measure of the thermodynamic availability loss in a heat exchanger, improved heat exchange is one means of improving overall system performance. Heat exchanger effectiveness for an evaporator or a condenser is an exponential function of the number of heat transfer units represented as UA, where U is a suitably averaged overall air side-based heat transfer coefficient and A is the corresponding heat transfer area; that is, E = EXP(UA) (Kays and London, 1964). Hence, increasing either the overall heat transfer coefficient (e.g., by increasing air flow, as the air side heat transfer resistance is liable to be controlling in most cases of practical interest) or the coil area will improve heat transfer.

Many manufacturers have been increasing the coil area in the last several years. According to West (1976), outdoor coil areas of some manufacturers' equipment have nearly doubled since 1973. Comparing for example General Electric's new high-efficiency Weatherton "Executive" heat pump with its earlier two-ton* heat pump, nearly 60% more coil face area per unit of nominal (cooling) capacity is provided with the newer model (Table 3-1).

Table 3-1. Trends in Electric Heat Pump Heat Exchanger Area as a Contributor to Higher COP (Gordian Associates, 1978)

	1976 Model	1977 Model
Nominal Cooling Capacity, Btu/hr	24,000	26,000
Outdoor Coil Face Area, ft^2	8.5	14.5
Face Area/Capacity Ratio, ft^2/Btu/hr x 10^4	3.5	5.6
Heating COP		
17°F	1.6	1.9
47°F	2.5	2.7
Cooling, EER, 95°F	7.0	8.1

Increasing the coil area obviously increases heat exchanger cost; however, there are other limitations to increasing the size of the heat exchangers. For example, there is decreased dehumidification capacity in the cooling mode, which results from a larger indoor coil and an attendant higher coil temperature. Increasing the outdoor coil (the condenser, in the cooling mode) requires an increased amount of refrigerant charge.

*1 ton = 11,950 Btu/hr.

Empirical data on system reliability shows a decline beyond a certain charge quantity, due, in part, to increased probability of liquid sludging. Hence, this approach has its restrictions. Increased air flow, i.e., coil face velocity, for both the indoor and outdoor coil is also a means of improving heat exchanger effectiveness through its effect on U, the overall heat transfer coefficient. Since the refrigerant side condensing or boiling heat transfer coefficient is much greater than on the air side,* the air side film coefficient is controlling, i.e., $U \simeq h_{air}$, the latter varying approximately as the eight-tenths power of the face air velocity, $h_{air} \sim u^{0.8}$. Fan power, however, increases as the cube of the air flow, and increased noise is also a consideration. If the coil in question is the indoor coil, chill draft condensate blowoff in the cooling mode can be a possible problem as well as the fact that most heap pump indoor coils already approach the ARI maximum of 37.5 cfm for each 100 Btu/hr.

Fan Efficiency and Power Requirements

Another major source of energy loss is outdoor and indoor fan power consumption; hence, decreasing fan power requirements will improve the COP. While necessary to provide air movement in a practical system, fan power represents another source of loss of thermodynamic availability in implementing an ideal Rankine cycle heat pump. Fan "air watts" are a product of heat exchanger pressure drop and volumetric air flowrate and thus varies as the cube of air velocity. Fan power is the product of the "air watts" times the motor efficiency. Heat transfer coefficient increases (which is one way of increasing coil effectiveness) must come at the expense of increased air velocity, which means increasing the air side pressure drop and the fan power requirements. On the other hand, it may be possible to maintain the normal capacity with a slower larger-diameter fan at considerable savings in power, as is illustrated in a comparison of the General Electric 1977 high-efficiency or "Executive" heat pump with its 1976 model (Table 3-2).

SYSTEM DESIGN

After efficiency of the basic heat pump engine has been attended to, recognizing the inherent thermodynamic problems of operating with a low-temperature ambient as a source thermal reservoir, proposed improvements in system design have generally been aimed at reducing the necessity of such a mode of operation. Usually this involves substituting a water

*Not all of the COP improvement show is attributable solely to increased coil area.

Table 3-2. Decreased Outdoor Fan Power as Contributor to Higher COP[a] (Gordian Associates, 1978)

	1976 Model (GE)	1977 Model (GE)
Nominal Cooling Capacity, Btu/hr	24,000	26,000
Outdoor Fan (propeller)[b]		
Diameter, in.	18	22
Motor horsepower	1/3	1/6
Motor speed, rpm	1075	825
Airflow at 0-in. wg, cfm	2600	2550
Outdoor Coil		
Face area, ft^2	8.5	14.5
Face velocity, ft/min	301	176
Heating COP		
17°F	1.6	1.9
47°F	2.5	2.7
Cooling EER, 95°F	7.0	8.1

[a]Not all COP improvement shown is attributable to decreased outdoor fan energy.
[b]No change in indoor fan, which is centrifugal, 9 x 9 in., 1075 rpm and has a one-quarter horsepower motor. Rated air flow, 800 cfm.

thermal reservoir as either an intermediary source or auxiliary supplement. Used as an intermediary, the water may be preheated as groundwater or through daily operation of a heat pump with an auxiliary energy source. The latter may include solar collectors. As an auxiliary supplement, a hot reservoir may be used to warm the air circulated through the residence while the heat pump is temporarily idle.

Water-Source Heat Pumps

Although most of the unitary equipment manufactured and installed in this country is of the air-to-air type, water-source heat pumps are growing in market importance because of their adaptability to water-based solar heating systems, to multizone or multiunit applications where energy recovery is a possibility, and because of some attractive features. Water-to-water heat pumps are common in applied or built-up heat pump installations where internal heat sources or heat or cold reclaim is possible.

First, water-source equipment offers higher COPs and a smaller variation of COP with source temperature, at least in principle, since the source temperature is maintained at 32°F or higher. If this has to be accomplished by resistance heating of the water, the overall COP will be no greater than one. The source of the heated water may be a lake, a river or wellwater (sources with obvious limitations in terms of availability, equipment cost, etc.); a solar or off-peak thermal storage system; or, in commercial or multiunit residential installations, a common water loop for thermal energy source and sink.

A second desirable feature is its compatibility with a water tank thermal storage system. This allows a water-source heat pump to be used in solar heating or cooling systems or to otherwise utilize thermal storage (for load management). Water-source heat pump evaporator coils do not experience frost problems (unless ice-making is a design feature of the system), a factor further contributing to their high seasonal efficiency.

The problems encountered in developing water exchangers for heat pumps center on the heat transfer limits of the exchanger. Carrier, for example, no longer has any water exchanger heat pumps. However, there are a few firms, mainly in southern states, that do market them. Climate Master heat pumps, for example, operate on wellwater as a reservoir with the water pumped to an exchanger constructed of intertwined coils. The water table in Florida is at 60°F year-round. This is considered necessary for the present heat pump models. With it, coefficients of heating performance are almost 3 all the time. These models can also achieve higher coefficients of performance for air conditioning so the energy operating cost saving over the long summer contributes further to their economy.

The market for water-source pumps was originally developed for the condominiums of Florida and is now considered very mature. It was desirable to supply each apartment owner with his own heat pump air conditioner unit so that utility costs could be determined on an individual basis. Using the common reservoir of wellwater circulated through all the systems, the heat pump operating electrically is maintained on an individual basis. The units are totally self-contained and need not expose any parts to the corrosive marine environment. A water-based unit, 50 WQ, once marketed by Carrier, was very similar in design and purpose. However, it was meant for closed loop use and would use a central cooler and boiler to maintain the reservoir. The water-based heat pump actually solved a problem with deterioration of exposed air-to-air exchangers that was present in both heat pumps and air conditioners.

Now that the technology has been established, it is being developed in the direction of utilizing water tables of lower temperature. Climate Master (now Friedrich of Wylain) expects the new models to operate on water down to 45°F, which would make them applicable up to the Canadian border. With such a capability, the use of a solar supplement in areas such as Minnesota, with a cold but sunny climate, would become economically feasible.

With the stability of a water-source temperature, such as that based on a water table, there is no change in heating-cooling capacity and COP with extremes in outside temperature. As a result, there is no requirement for the fairly elaborate measures described above to provide some means of adjusting compressor system capacity to the demand. Extremes of temperature will only occasion longer "on" cycles.

It is currently expected that in late 1980 Wylain will have a unit capable of operating in a 45°F water source; however, the danger of freezing is very real as experienced, for example, in the eastern Maryland experiment described in Chapter 5. One of the most essential safety features required will be reliable detection of refrigerant flow stoppage under these conditions. In that situation, immediate remedial measures must be taken.

Solar-Assisted Heat Pump Systems

It has been pointed out by Drucker et al. (1975, 1976) that a water-based heat pump can be supplemented with solar energy at low capital cost because required operating water temperatures can be kept fairly low. In addition, the coefficient of performance of water-to-air heat pumps can be fairly high because one does not have to heat the air exchanger to evaporate frost buildup. This is, of course, a major depressant of COP, as temperature hovers in the 32–40°F range.

In solar heating and cooling system applications, water-source heat pumps suffer from the limitation that condenser heat on the cooling cycle must be rejected to water or, in the absence of a water-based heat sink, to air via an air-cooled heat exchanger. This may involve additional fan or pumping power, diminishing the COP for cooling.

A dual-source (water and air) heat pump circumvents these disadvantages in that the heating cycle evaporator heat exchanger can be immersed in water or air. This enables the water loop to be used during at least that part of the heating cycle where the water temperature exceeds that of the ambient air, and the air coil to be used in warm heating weather or for cooling operation. Some of Wylain's developments toward a dual-source unit with daylight air operation and automatic shift at night to water source, at present limited to 60°F, have been described. However, this does not bar eventual operation at 45°F. This will remove unnecessary daytime thermal depletion of the water reservoir.

The basic water-to-air heat pump assumed in the work by Drucker et al. (1975) is shown in Figure 3-1. This system is assumed to be used in a large complex, and its operating cost is modeled for an entire year using comprehensive daily weather detail such as wind, rain, clouds and humidity, in addition to temperature. Typical building type is a school,

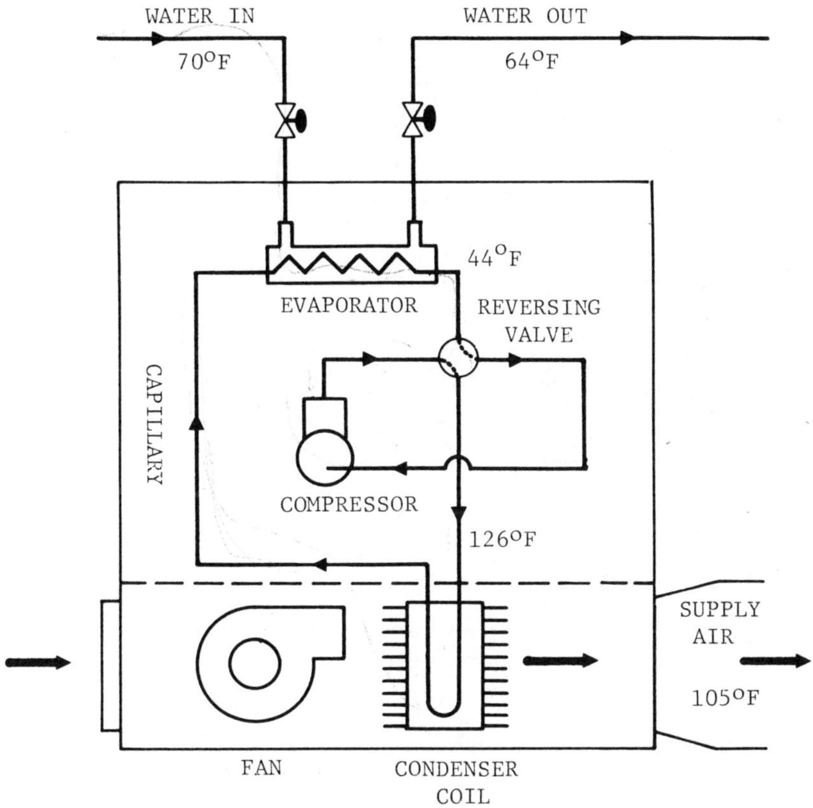

Figure 3-1. Water-to-air type heat pump, heating cycle (Drucker et al., 1975).

a schematic of which is shown in Figure 3-2. The model was run for schools with floor areas of 7920 m^2 in Alexandria, Virginia, and Minneapolis, Minnesota, which are operating with heat pumps. The choice of a multiplicity of unitary heat pumps has already been demonstrated to offer better energy conservation potential than central systems in conventional buildings. Similar to separately controlled resistance heaters, for example, that in Minneapolis, with 1400 ft^2 of solar collector added, city of switching on/off and heating/cooling. They were able to show, for example, that in Minneapolis, which 1400 ft^2 of solar collector added, they could save about 600,000 kWh annually out of supplemental heat that the school is currently forced to supply. Similar results were obtained on an office building with 10750 m^2 floor area.

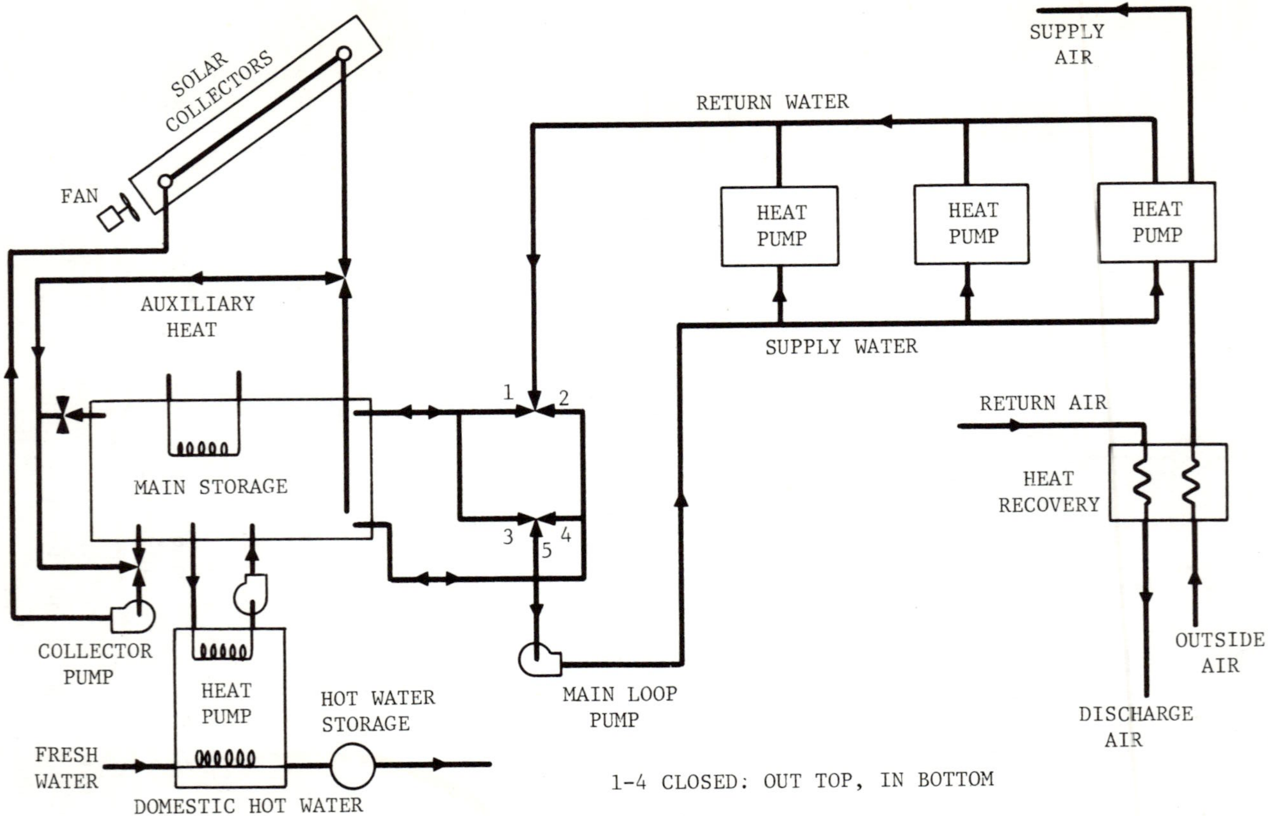

Figure 3-2. Solar-assisted, closed-loop heat pump system (Drucker et al., 1975).

Thermal Reservoirs

Use of thermal reservoirs in heat pump systems is being actively explored in a number of the utility experimental programs described in Chapter 5. The reservoirs are generally of two types: well-insulated large hot water tanks for uses with temperatures up to 190°F (88°C); and ceramic aggregates (hot rocks) for use up to 1400°F (760°C). In both cases there is provision for circulating air and/or water through an immersed heat exchanger coil and for heating the reservoir by electric resistance. Essentially commercially available electric hot water tanks and large electric muffle furnaces serve well as the principal structure for the reservoir.

Water reservoirs are very popular in the solar assist experiments. They are also useful in a few programs where ice storage is used during the summer to improve air conditioning operation, but such storage is rather controversial because its capital costs can be very high. The utility point of view is that it can serve well to balance their load, using off-peak electricity for its heat. So far, however, there is one promising avenue that has not been explored. For several hundred dollars, a well-insulated 125-gallon electric water heater can be purchased and modified so that its water can be heated by the heat pump. It seems feasible, using a well-programmed control module such as the Yorkguard (see Chapter 2), to charge it with a heat pump operating continuously during a diurnal winter high. The home heating unit can then work off the reservoir to stabilize indoor temperature with the heat pump activated to heat the reservoir at low outside ambient temperature only when the reservoir temperature falls very low, such as 70°F. Under severely low ambient temperatures, the built-in electric supplement could be activated. This mode of operation would probably only be economical in the U.S. Northeast; current system economics for the south indicate it would be a poor investment.

GAS-FIRED HEAT PUMP ENGINES

The direct use of gas instead of electricity promises to raise the basic effectiveness of heat pumps with regard to utilization of the directly realized energy of combustion, two-thirds of which is now lost in electricity generation. Using a gas-fired prime mover as a power source offers the use of engine waste heat rejected in coolant or exhaust gas to supplement the refrigeration cycle output in the heating mode, thus substantially increasing the overall onsite system COP. On the other hand, the extra heat produced in close proximity to the cooling heat exchanger becomes a liability in the cooling mode. As in electricity generation, this heat is dissipated; however, the electrically powered cooling engine need not contend with nearby exhaust heat.

The possibility of strong gain in the heating mode has motivated a great deal of interest both in DOE programs and privately sponsored research. This latter component indicates how realistic are the prospects for near-term commercial introduction of gas-fired heat pumps. The new concepts being developed are mainly based on the use of combustion to drive a thermal engine rather than an absorption cycle.

The general aspects concerning advantages and disadvantages of gas-fired thermal engine heat pumps have been described by Auxer (1977), Swenson (1977), Fallon and Iles (1978), Lawrence (1978), Colosimo (1978), and Liebowitz and Colosimo (1978), who have all proposed various thermal engine configurations for use as heat pumps.

Depending on the particular prime mover or refrigeration cycle contemplated, thermal engine heat pumps have the following advantages over electric heat pumps:

- high overall efficiency (in a primary energy sense) in comparison to state-of-the-art electric heat pumps in heating mode and gas or oil furnaces;
- ability to utilize prime mover waste heat to supplement refrigeration cycle on heating mode;
- capability to modulate heat pump capacity by varying fuel input rate to prime mover;
- availability of onsite waste heat for evaporator defrost;
- adaptability to total energy systems or for operation independent of utility lines.

The major inducement to thermal engine heat pump development is the high combination engine efficiency and refrigeration cycle coefficient of performance obtainable in principle. Waste heat utilization onsite, facilitated defrost and ability to modulate capacity all add to high seasonal performance in the heating mode. Although electric heat pumps have certain advantages in the cooling mode, it is possible to design a thermal engine heat pump to "hold its own" on cooling as well.

Prime movers, especially turbomachinery, are relatively easily speed-modulated by controlling fuel flow to the combustor or bypassing some of the power fluid around the prime mover. This allows easier modulation of heat pump capacity than with the synchronous A.C. motors used in electric heat pumps.

Having a source of waste heat onsite not only makes supplemental heat available for augmentation of the output but allows its use for efficient evaporator coil defrosting in the heating mode. This could eliminate a major reliability problem as well as a cause of some efficiency loss.

The disadvantages of thermal engine heat pumps are listed below. These apply to varying degrees, depending on the particular system being considered:

- necessity of rejecting proximally generated prime mover waste heat in the cooling mode;
- greater system complexity which could result in potentially higher system cost and lower reliability;
- difficulty in designing a totally hermetic refrigeration system;
- aggravated noise and pollution problems due to onsite prime mover operation and fuel combustion; and
- much lower service lifetime under routinely interrupted operation.

Experience with small residential and commercial HVAC and refrigeration systems has shown that hermetic systems are considered more reliable. With an onsite prime mover, design of a hermetic refrigerant loop becomes difficult, although not impossible. Rotating high-speed machinery or unbalanced engine-compressors are bound to increase noise and vibration levels over hermetic electric motor compressors or gas furnaces. Much of this noise can be eliminated in a closed-system design, and with proper vibration damping and sound insulation, but at some increased cost. One also cannot neglect the greater system complexity compared to an electric heat pump. In general, practical thermal engine heat pump designs must have:

- two different working fluids, requiring their sealing and isolation from each other;
- many moving machined surfaces requiring machining to close tolerances and lubrication (most heat pump concepts use refrigerant oil, but the engine lubrication may be dry);
- several thrust and journal bearings;
- feed pumps or blowers, engine-coolant pumps;
- engine and heat pump control devices;
- a separate starting system in some cases; and
- materials to withstand high static or dynamic operating stresses at high temperature.

This increased complexity is likely to result in high first-cost equipment and potential problems of system and component reliability, especially in a product that is expected to operate essentially maintenance-free for a long time. The major problems of the prime mover coupled in the heat pumps are found in its susceptibility to failure or accelerated breakdown in interrupted service. When electricity is generated at a central plant in almost continuous operation, the interrupted operational requirements at the distant working end are then handled by the electric motors, which can handle it so much more appropriately. This is why the most immediate prospects of success for gas-fired heat pumps are being sought in large-scale commercial or industrial service. A unit suitable for residential use is generally considered at least a decade away.

Most of the experimental units being developed are based on a combination of cycles, one of them being the Rankine cycle, based on pressure-temperature paths through partially saturated and superheated vapor phases.

Stirling-Rankine Heat Pump

An example of a Stirling cycle engine driving a Rankine cycle heat pump is the system under development by General Electric Company's Space Division. Its development has been described by Auxer (1977) and some of its performance characteristics by Richards and Auxer (1978). The prime mover is a free piston Stirling engine originally developed for the American Gas Association by Sunpower Incorporated (Beale et al., 1973). It provides linear reciprocating motion to drive a double-acting, inertia-type Rankine cycle vapor compressor. This permits complete isolation of the Stirling engine working fluid from the ambient and from the refrigeration loop without the use of dynamic seals.

The engine/compressor assembly is shown in Figure 3-3, the left half of which shows the Stirling engine. It has a single cylinder, with the displacer and power piston contained within the cylinder in an in-line configuration. The engine housing assembly is the pressure vessel member and contains the heater head, regenerators and cooler subassemblies. The energy source for the engine is the combustion of a premixed air/natural gas, using a radiation-cooled transpiration burner, as shown in Figure 3-4. Mass flowrates of m = 37.3 lb/hr of air and 1.94 lb/hr of gas are indicated with excess air of 15% for combustion. The liberated energy is transferred to the Stirling engine working fluid, helium, through the finned tube heat exchanger to about 1000°F (536°C) without preignition. Temperature of the exhaust products from the heat exchanger is about 1300°F and can be used to preheat the combustion air. Operating results for the combustor are shown in Table 3-3. A mixture of ethylene glycol and water is used in a hydronic loop to transfer engine waste heat to the heat rejection zones appropriate for heating and cooling, using a three-way valve.

The inertia compressor assembly for the heat pump is a two-cylinder, two-piston, single-acting unit shown on the right side of Figure 3-3. The cylinders are configured in-line as mirror images of each other about the mid plane. Cylinder assemblies are fabricated from aluminum with a cast iron sleeve. Each cylinder is instrumental with a pressure transducer. The

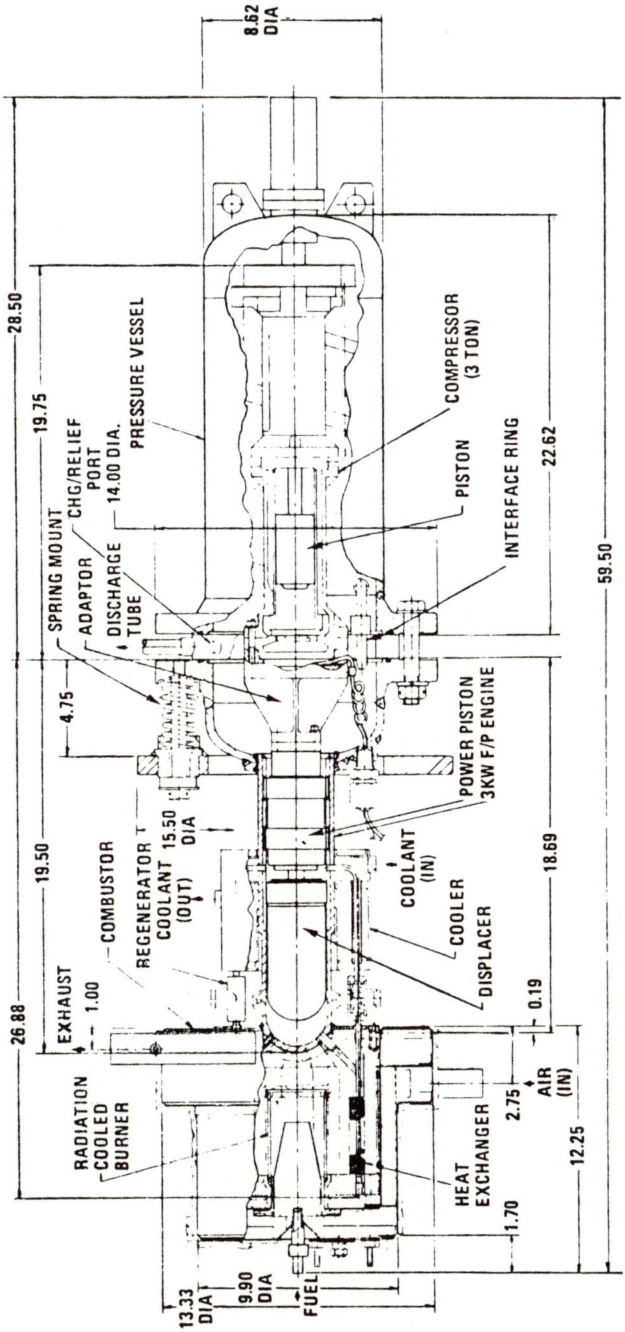

Figure 3.3 Engine compressor assembly for gas-fired heat pump (Richards and Auxer, 1978).

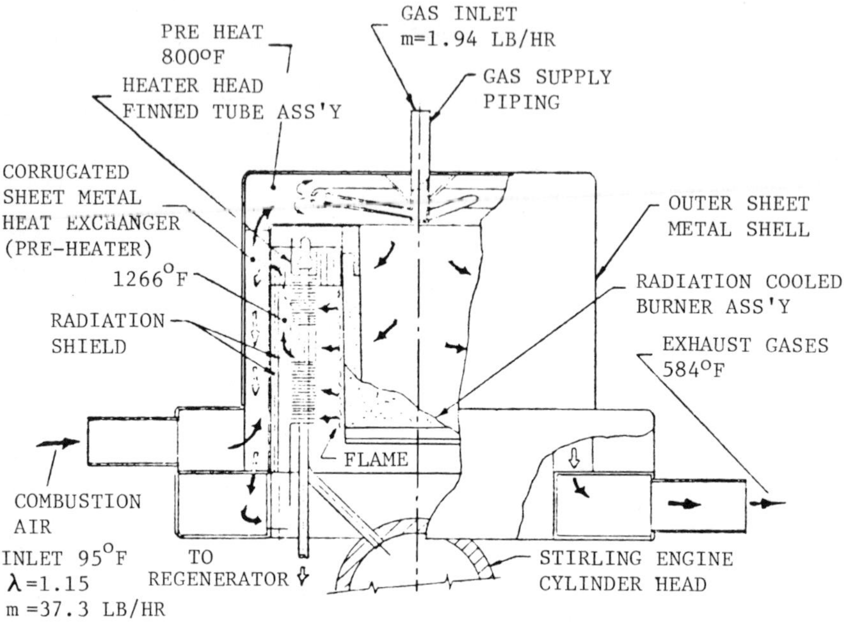

Figure 3-4. Stirling engine combustor (Richards and Auxer, 1978).

cycle controlling valves rest in the cylinder lines and separate suction refrigerant from discharge refrigerant. The pistons are jointed together with a connecting rod and close-fitting wrist pins. The volumes contained between the valves and the outer faces of the pistons are the compressor working volumes. The volumes between the midhousing plate and the connecting end side of the piston comprise the gas spring volumes.

Table 3-3. Combustor Design Point, Definition and Operating Results (Richards and Auxer, 1978)

	Predicted Design Point Definition	Test Results
Burner Energy Release, Btu/hr	42,000	42,000
Effective Combustor Efficiency, %	80	85
Air Preheat, °F	800	815
Excess Air, %	15	15
CO Emission Level, %	0.04	0.04
Air Side Pressure Differential, in. of H_2O	1.0	1.0

Checkout testing of subsystem performance indicates that predicted performance has been realistic as shown in Table 3-4. In private communication, William Auxer of General Electric's Space Division at King of Prussia, Pennsylvania, has indicated that this development is primarily aimed at continuous industrial use. To avoid interrupted operation, General Electric is experimenting with multiple steps in capacity. Auxer speculates that it will be at least the mid-1980s before they will have a good idea about the commercial viability of this approach.

Table 3-4. Stirling-Rankine Subsystem Performance (Richards and Auxer, 1978)

FPSE Performance Predictions and Test Conditions

Parameters	Predicted Values	Test Conditions
Charge Pressure	60-80	40-80
Firing Rate, kBtu/hr	40	20-40
Frequency, Hz	27	18-26
Power Piston Stroke, in.	1.7 max.	0.9-1.4
Displacer/Piston Phase Angle, degrees	30 nominal	55-75
Output Power, W	1800	1400

Free Piston Stirling Engine Design Parameters

Working Fluid	Helium
Charge Pressure, psia	870/1160
Power Level, kW	3.0
Design Efficiency, %	32
Heater Head Temperature Level, °F	1200
Coolant Inlet Temperature Level, °F	140
Power Piston Stroke, in.	1.7
Frequency, Hz	26

Subsystem Performances Measured and Predicted

Subsystem	Product Goals	Prototype 1 Test Results	Prototype 2 Prediction
Combustor	80%	84%	85%
Engine	32%	30%	35%
Refrigeration Loop COP	3.5	3.2	3.5

Rankine-Rankine Heat Pump

At Mechanical Technology, Inc. in Ballston Spa, New York, there is an extensive heat pump development program underway emphasizing gas or oil firing. Practically all of it is privately funded, as is all of their work for residential applications. The industrial program includes electric drive and is oriented toward waste heat recovery by raising the temperature of exhaust gases. Only the developments by Swenson and Rose (1977) on a steam-Rankine heat pump combustion are fully in the public domain. Their publication describes mostly developmental prospects. However, the system was, in fact, eventually built. Whereas General Electric collaborated with DOE and the American Gas Association, Mechanical Technology Incorporated was sponsored by ERDA (predecessor federal agency for this work) and Consolidated Natural Gas.

The MTI steam-Rankine system emphasizes the use of rotating machinery, by contrast to the GE linear reciprocal machinery. A steam generator drives a turbine on a common shaft with a small alternator and a centrifugal compressor. Its scheme in heating and cooling modes is shown in Figures 3-5 and 3-6, respectively. Major components of the system are listed in Table 3-5. Fluid states in Figure 3-6 follow the same code indicated in Figure 3-5.

The reversible refrigeration loop is similar to that employed in electric heat pumps. The expansion valve (XV) and reversing valves (four-way) are shown. Steam is delivered to the turbine at 21 kg/cm^2 Abs (300 psia) and 590°C (1100°F). A single reheat and regeneration is employed. The operation of the system will employ several important features, as follows:

1. Phased heating of the conditioned air. First, the pumped heat is transferred to the return conditioned air. Secondly, the steam turbine's exhaust is used to heat the air. This significantly reduces the head pressure requirement of the heat pump compressor. Second-law considerations show that reducing the work per unit heat being pumped is a favorable trade-off for the resulting elevation in the turbine's backpressure.
2. Direct heating of conditioned air with steam, bypassing the turbine at very low outdoor temperatures. The bypassed steam heats the conditioned air, again after it is heated by pumped heat. By this method, useful heat pumping can *coexist* with heating by means of bypassed steam. This technique, coupled with that of item #1 (above) will lead to useful heat pumping down to about -15°C with single-stage centrifugal compression.
3. Fully modulated steam cycle operation. Full utilization of this technique can only be accomplished when both the heat pump compressor and the air-moving fans are speed-coupled to the prime mover. An advantage in this approach is that at low-capacity operation, air-moving power requirements do not destroy the basic energy-saving

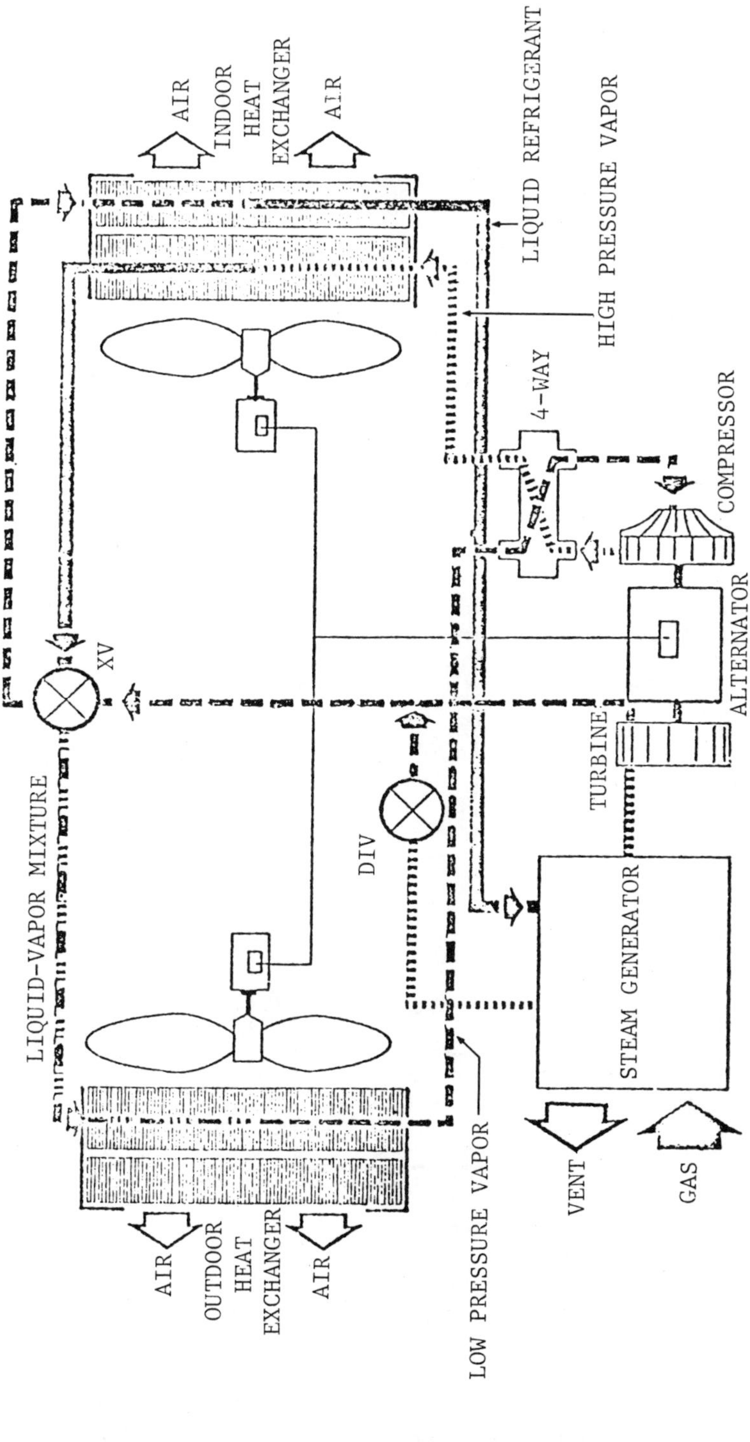

Figure 3-5. MTI gas heat pump system, heating (Swenson and Rose, 1977).

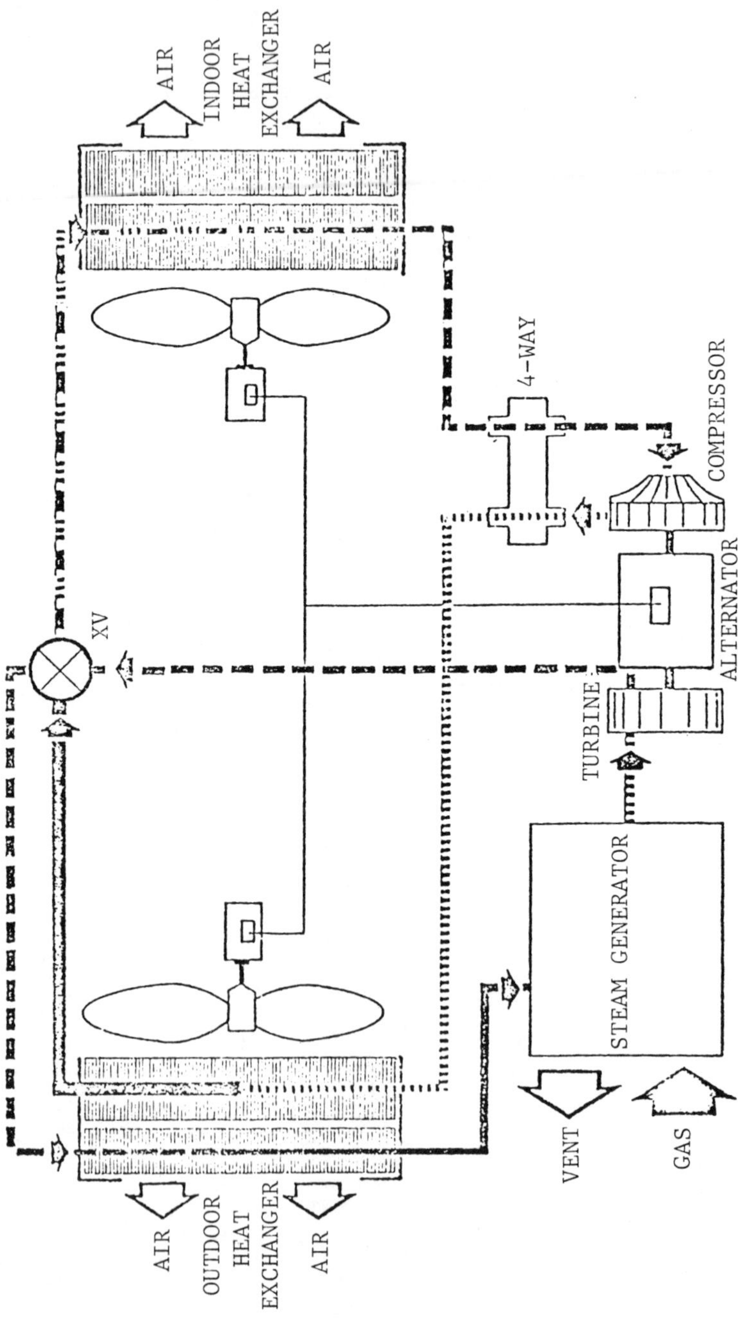

Figure 3-6. MTI gas heat pump system, cooling (Swenson and Rose, 1977).

Table 3-5. Major Components of MTI Steam-Rankine Heat Pump (Swenson and Rose, 1977)

Steam Generator	Monotube, power burner with integrated superheater, reheater and combustion air preheater
Turbine	Single stage, axial flow with reentry
Alternator	Three-phase inductor homopolar
Bearings	Refrigerant gas-lubricated
Breadboard	Pivoted shoe
Product	Hydresil® foil-type
Shaft Seal	
Breadboard	Combination floating labyrinth and carbon-ring face-seal
Product	Hermetic
Compressor	Single-stage centrifugal, fixed geometry
Air/Stream and Air/Refrigerant Heat Exchangers	Flat tube and plate-fin, all parallel circuits Series air flow
Feed Pump	Hermetic diaphragm-type with gear-type booster
Drive Motors	
Breadboard	dc
Product	ac squirrel cage

advantages of modulation. The only control inputs are to the burner and to the turbine throttle.

4. Phased cooling of system condensers (cooling operation). This is the second law analog of item #1, except as employed during cooling operation. During cooling, heat being rejected to the surroundings from the heat pump (air conditioning) circuit is given first access to cooling air.

The major drawback to this system lies in the fact that the refrigeration loop is not hermetic, so that a dynamic shaft seal is necessary on the common shaft linking steam turbine to the alternator and refrigerant compressor. Air filtration can cause corrosion on the steam side and reduce heat exchanger overall heat transfer rates on the refrigerant side and, hence, the capacity. This system is designed to use R-11 as the refrigerant, which has been tentatively identified in connection with its possible reactivity with atmospheric ozone. This will not be a problem if sealing can be near perfect.

The MTI system has been specifically designed for commercial use in the north central U.S., where heating requirements are high and ambient temperatures are most unfavorable for heat pump operation. Modeling the performance of their system for these conditions, Swenson and Rose arrive at coefficients of performance (based on prime mover energy) of 1.2 at 14°F (-10°C) for estimated fuel savings from a 100% efficient gas-fired furnace of about 17%. This may not be enough to make the unit economical, considering the winter night-time lows, unless it is adjoined to a diurnal thermal storage system. The storage will also help in relieving the system duty cycle of having to provide temperature stabilization directly.

In a related development at MTI, described by Williams (1978), an industrial heat pump is being designed to provide process steam from waste heat. Their plans are aimed at installation in a Celanese Fiber Company demonstration plant in 1980. The concept of a heat pump for such applications is illustrated in Figure 3-7.

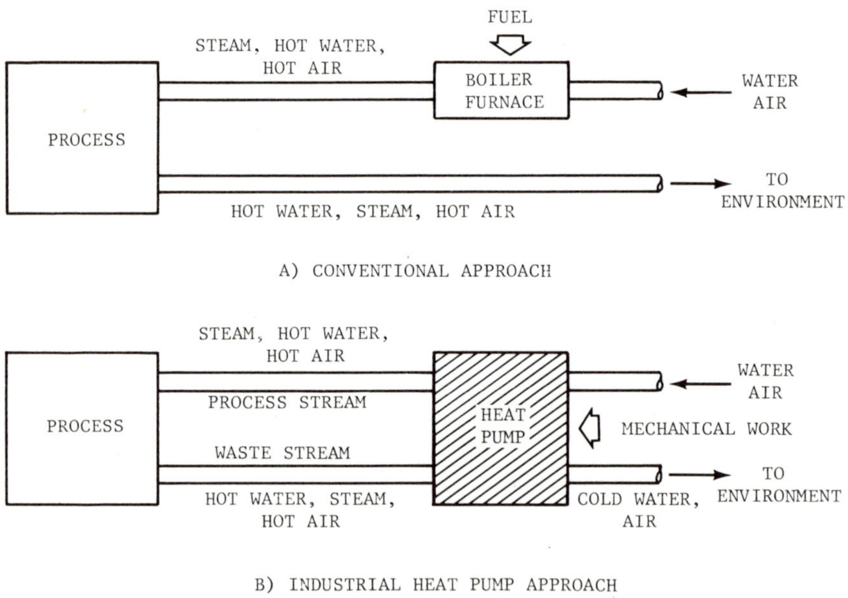

Figure 3-7. Heat pump use for process heat (Williams, 1978).

The MTI heat pump uses an organic Rankine cycle to apply a portion of the heat of the waste stream in driving a heat pump compressor, which extracts the remainder of the heat from the waste stream. A simplified schematic of this concept is shown in Figure 3-8.

The power (or bottoming) cycle of this system uses an organic working fluid R-11, with turbine inlet conditions of 160°F (71.1°C) and 60.45 psia

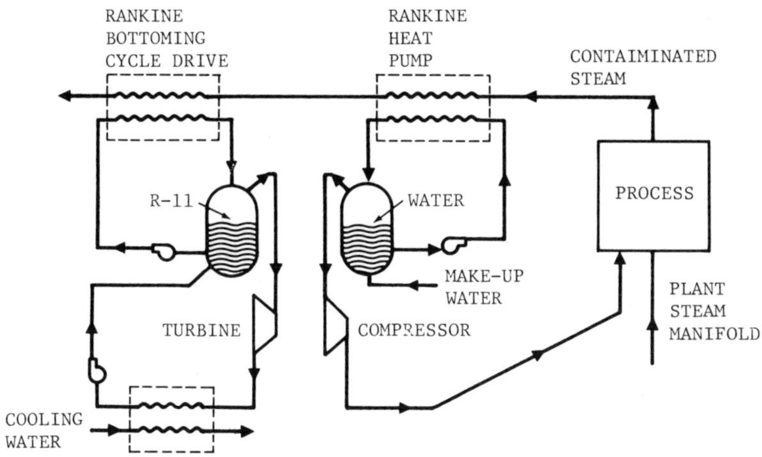

Figure 3-8. MTI industrial heat pump for process steam (Williams, 1978).

(4.25 kg/cm^2). The heat pump cycle uses steam as a working fluid in an open-loop reverse Rankine cycle. Heat pump evaporator conditions are 190°F (87.8°C) and 9.4 psia (0.66 kg/cm^2). The heat pump condenser is the process itself. The equivalent condensing conditions are 228°F (108.9°C) and 20 psia (1.40 kg/cm^2), with steam being superheated at the compressor exit to 349°F (176°C).

This combination of an organic Rankine power cycle driving an open-loop steam Rankine heat pump cycle recovers approximately 16.6 x 10^6 Btu/hr (4864 kW) from a waste stream flow of approximately 26.8 x 10^6 Btu/hr (7852 kW), with virtually no external energy source other than the waste stream itself.

Brayton-Rankine Combination

As a counterpart to steam turbine use, a gas-turbine Rankine heat pump is being developed by Garrett-AiResearch Manufacturing Company of California under the sponsorship of the American Gas Association (AGA), which has been considering both closed and semiclosed Brayton cycle units looking toward a nominal cooling capacity of 120,000 Btu/hr (10 tons) and a heating capacity of about 180,000 Btu/hr at maximum turbine speed of 90,000 rpm.

One prototype Brayton-Rankine heat pump configuration is shown in Figure 3-9. The system employs a subatmospheric open-cycle gas turbine with a centrifugal compressor for compressing the exhaust gas before

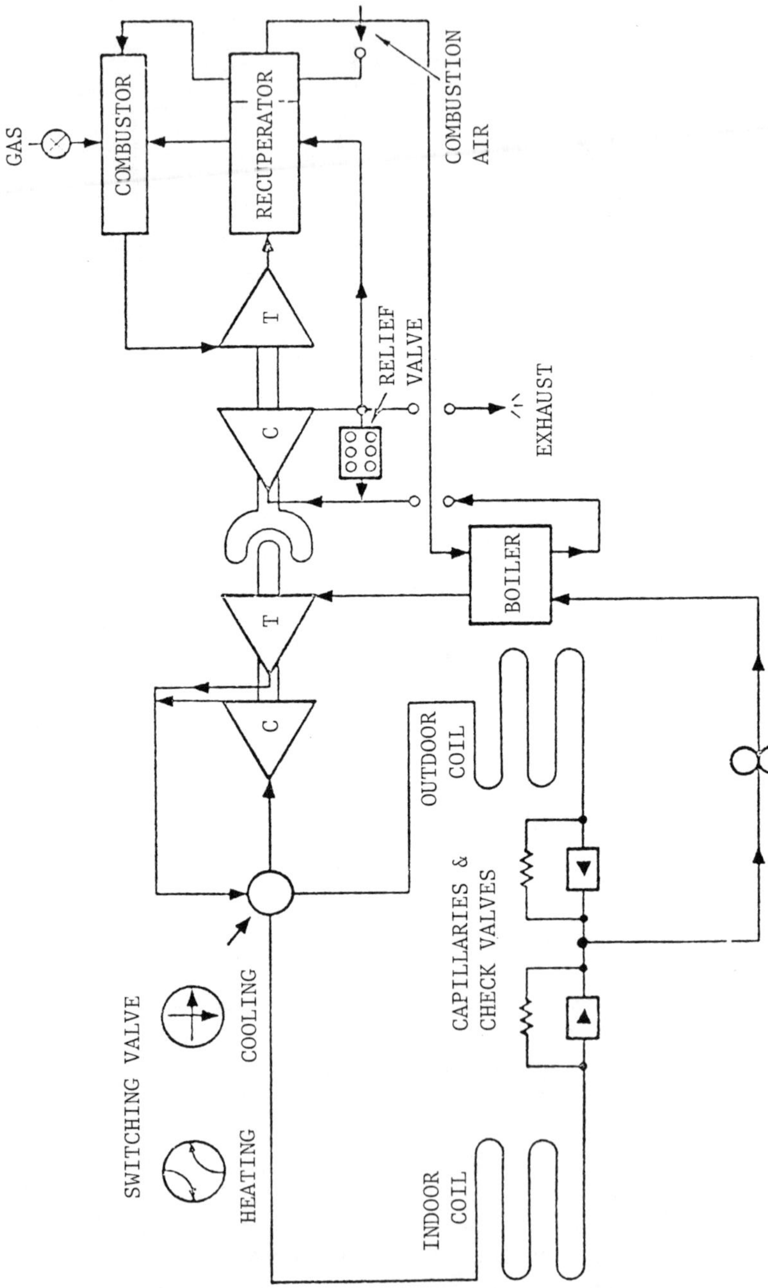

SOURCE: Garrett Corp. 1976

Figure 3-9. Brayton-Rankine heat pump (Garrett Corp., 1976).

rejection to the atmosphere. The prime mover shaft is connected via samarium-cobalt magnetic coupling to the centrifugal-type refrigerant compressor. Thus, the refrigerant loop is completely hermetic. The Brayton rotating group employs air-lubricated foil journal and thrust bearings. Other features of the system are engine and exhaust heat recovery and the use of a refrigerant boiler and expander as the startup system for the engine. Note that the diagram shows an additional turbine running on refrigerant vaporized in an R-12 boiler by gas combustor exhaust heat. This is used as a bottoming cycle to increase efficiency and serve as a startup mechanism.

Garrett is developing the Brayton-Rankine heat pump as a unitary gas heat pump for commercial applications. While the design is amenable to scaleup, according to the developers, it is unlikely that a smaller unit than the prototype 10-ton size could be developed at a reasonable first cost. A current development can provide heat for drying milk; it has been described by Fallon and Iles (1978). A typical milk dryer installation (Figure 3-10) would be replaced by a subatmospheric Brayton heat pump installation (Figure 3-11). The system would meet the demands posed by the typical milk dryer parameters (Table 3-6).

The milk is initially condensed in steam-heated evaporators from about 8% solids to 45% solids. The water vapor is condensed and leaves the

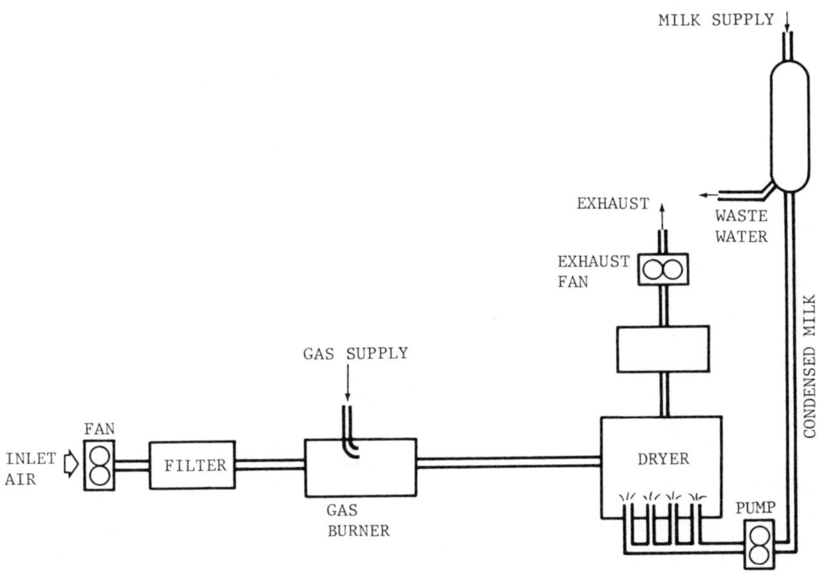

Figure 3-10. Typical milk dryer installation (Fallon and Iles, 1978).

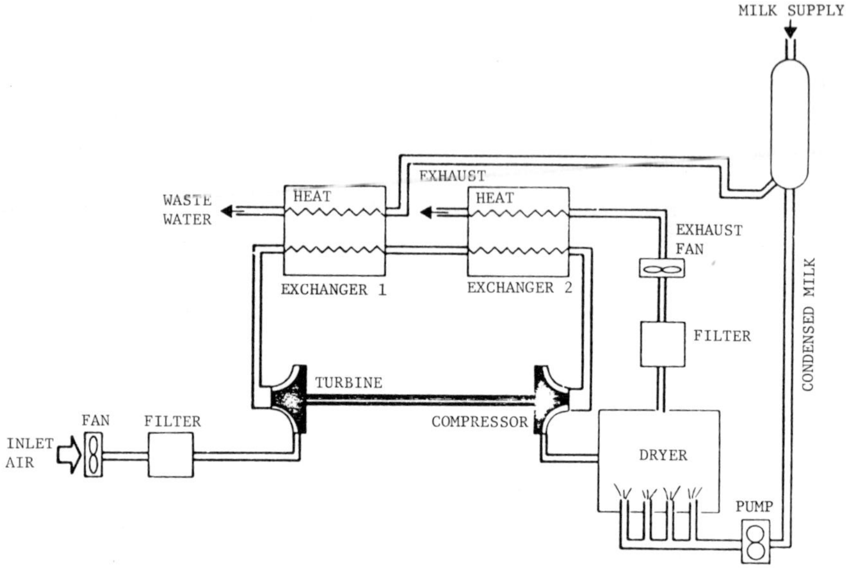

Figure 3-11. Milk dryer with subatmospheric cycle heat pump (Fallon and Iles, 1978).

Table 3-6. Typical Milk Dryer Facility Parameters (Fallon and Iles, 1978)

Inlet Milk Supply	67,000 lb/hr
Air Flow	3,200 lb/min
Dryer Inlet Temperature	325°F
Dryer Exhaust Temperature	190°F
Evaporator Wastewater Flow	860 lb/min
Evaporator Wastewater Temperature	140°F
Milk Powder Product	3800 lb/hr
Evaporated Milk Product	7500 lb/hr

process at 120-160°F. A portion of the condensed milk is normally used directly in other products. The portion of the milk to be dried is sprayed through nozzles into the dryer, where the milk droplets meet hot, dry, filtered air and the solids content increases to at least 97%. All milk drying facilities also include refrigeration for storage of incoming milk and processed milk products, since any curtailment in the fuel supply or any plant failure could otherwise lead to spoilage. The object of this program is to recover the waste energy from the process, in particular, from the exhaust gas leaving the dryer, and to use it to reduce system energy requirements.

In the subatmospheric Brayton-cycle heat pump, ambient air enters the turbine after being filtered. Adiabatic expansion of the air through the turbine provides some of the power required to drive the system; the remainder is provided by an electric motor, as for the positive-pressure cycle. The air cooled by this expansion is first heated by the moist dryer exhaust gas in heat exchanger 2. The heated air is then compressed in the compressor to ambient pressure. This compression process further increases the air temperature to the desired value before it enters the dryer. System COP for this configuration is 3.4 for an ambient temperature of 65°F.

ABSORPTION HEAT PUMP CONCEPTS

Thermal engine heat pumps treated in the foregoing section have utilized homogeneous fluids (refrigerants) undergoing a Rankine or other refrigeration cycle, which are driven mechanically. It is also possible to pump heat by one of several absorption refrigeration cycles.

The following analogy exists between an absorption refrigeration cycle and a vapor compression, or Rankine, cycle. In the latter case, a compressor is used to increase the pressure of the refrigerant vapor prior to its condensation; in the former, the *chemical affinity* between a refrigerant vapor, an absorbent solution and a solution pump produces a similar effect. Indeed, the expression for the COP of a simple ideal absorption cycle has the form of a product of the efficiency of the Carnot engine cycle and the COP of a Carnot refrigeration cycle operating at the same limiting temperatures (Curran, 1975). Absorption air conditioning machines are commercially available in large-tonnage (100 and up) sizes. On the other hand, some unitary smaller tonnage residential and commercial absorption cycle air conditioners are also being produced by a single manufacturer, but to a steadily shrinking share of the total unitary air conditioning market. No absorption cycle unitary heat pump is yet commercially available.

Absorption cycle heat pumps, depending somewhat on the cycle and the particular design configuration employed, have the following major advantages:

1. They are inherently simple and potentially highly reliable. The solution pump and heat transfer circulation equipment (fan or pump) are the only moving parts. No compressor is needed.
2. They are highly efficient in the heating mode due to possible heat recovery from onsite combustion as supplementary heat to that provided by the absorption cycle condenser itself.
3. They have lower parasitic energy requirements than thermal engine heat pumps.
4. They use refrigerant-absorbent fluids, which are chemically nonreactive with atmospheric ozone.

5. There is easier coupling of absorption cycle generator to solar thermal source.
6. There is available onsite waste heat for evaporator defrost.

The following are the substantial disadvantages:

1. Comparatively large heat exchanger areas (at attendant high first cost) are required to realize acceptable COPs.
2. Usually they have lower cooling mode COPs than obtainable with Rankine refrigeration machinery (including electric heat pumps).
3. Toxicity of ammonia requirement for avoidance of direct refrigerant/conditioned air contact in aqua-ammonia systems.
4. Aqua-ammonia systems are corrosive so that steel, rather than aluminum or copper, must be used in construction.
5. There is a diameter limitation of generator pressure vessel leading to a tall unit for adequate vapor capacity.
6. The "plumbing" of absorption machines is apparently more complex.

Current well-known absorption air conditioning fluids have not been utilized successfully in absorption heat pumps because of inherent limitations in the fluid properties. However, the American Gas Association and Allied Chemical have identified a promising fluorocarbon refrigerant and organic absorbent pair, which they have maintained as proprietary. Their plans for a heat pump have been outlined by Colosimo (1978). A pressure-temperature-concentration plot for the organic pair is illustrated in Figure 3-12. The temperature and concentration points are accurate, but the pressure levels are somewhat distorted for illustration purposes. The numbered points refer to the fluid conditions at various locations in the hardware loop, shown in Figure 3-13. The organic pair has a number of advantages over conventional absorption fluid pairs. The first is that the organic fluid pair is compatible with aluminum, which means a less costly device. Original cost estimates suggest that the organic pair system could be as much as 25% less expensive to manufacture than an ammonia water system. Another advantage is that the slope of the constant concentration lines are more horizontal than the other fluids, which permits more heat transfer between various streams, as shown in Figure 3-12. The unit has an absorber heat exchanger, a generator heat exchanger, a precooler heat exchanger, and a separator or heat exchanger. All these internal heat exchangers contribute to this system's overall higher performance. Finally, the fluids are neither toxic nor flammable.

The hardware schematic for the absorption heat pump is shown in Figure 3-13. There are two circulating water/glycol loops. One, a cold water loop, always goes through the refrigerant evaporator and then either to the indoor coil for cooling or the outside coil for heating. The other,

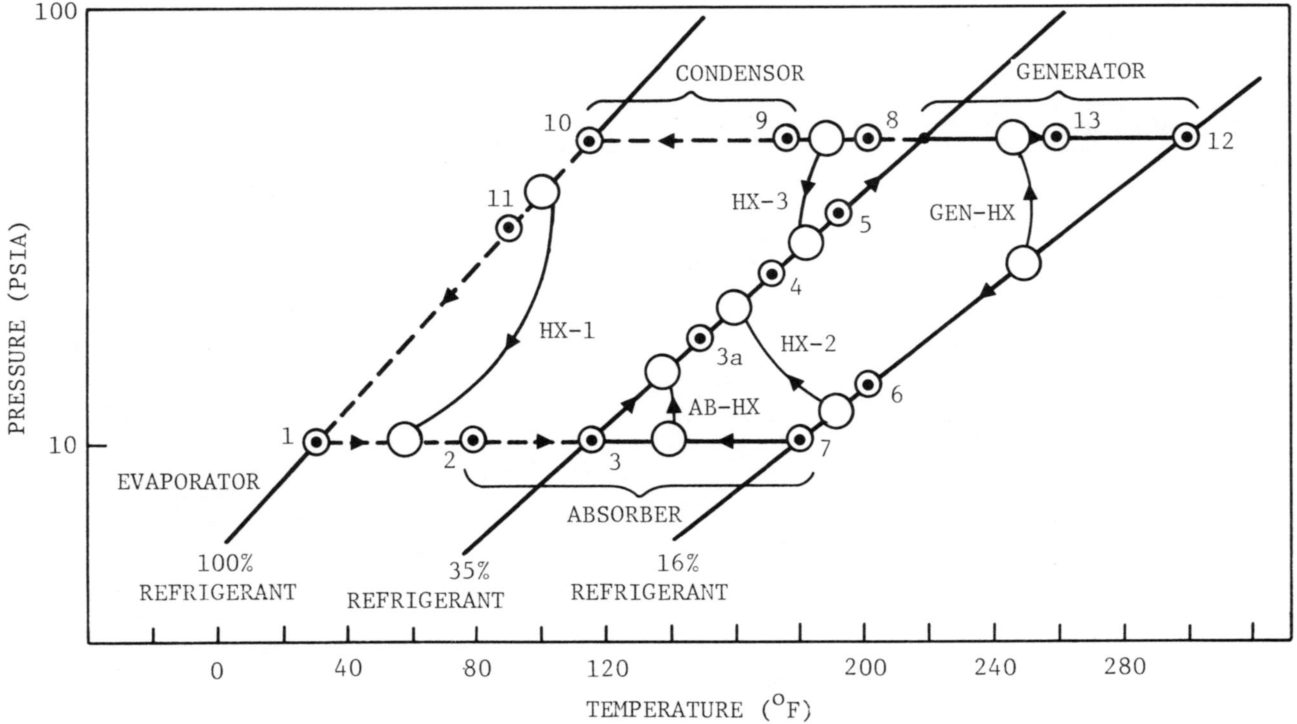

Figure 3-12. Idealized organic pair absorption cycle heating mode at 45°F ambient (Colosimo, 1978).

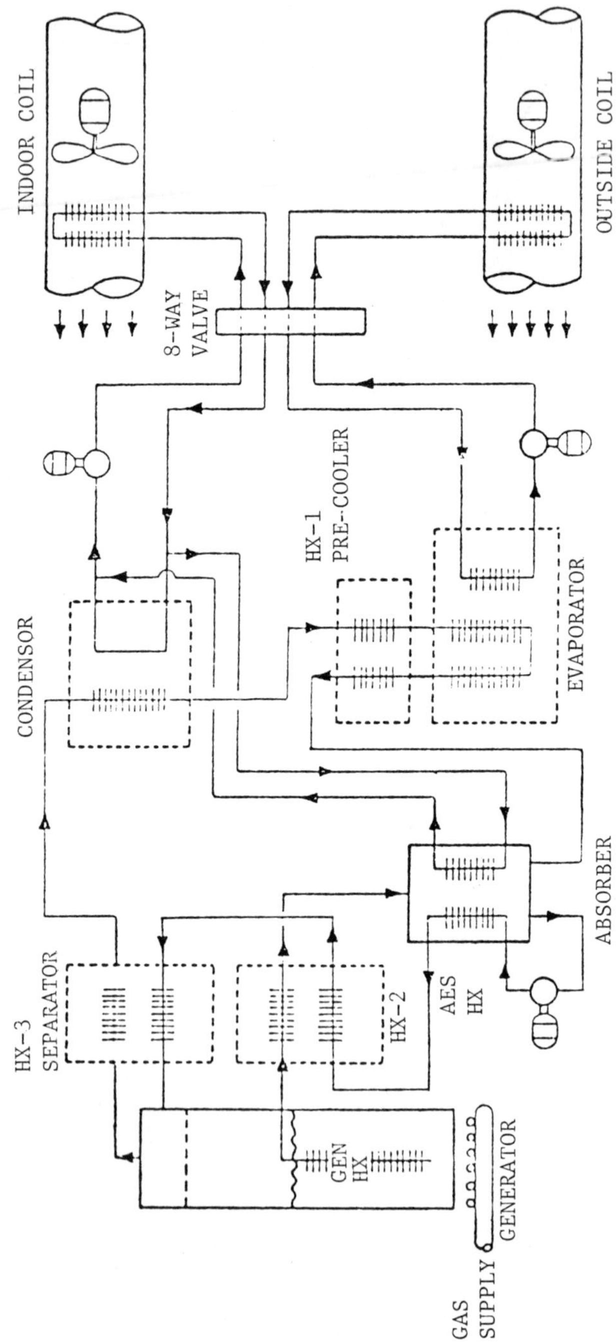

Figure 3-13. Organic fluid absorption cycle heat pump, heating mode (Colosimo, 1978).

a hot water loop, always goes through the absorber and refrigerant condensor, then to the outside for cooling or the indoor coil for heating. This transfer of the two water/glycol streams requires an eight-way valve. Although only a single heat exchanger is shown on the indoor coil, a double heat exchanger will probably be used like the other heat pump system. During the cooling mode, the water will go through only one of the heat exchangers; during the heating mode, the water will go through both heat exchangers to prevent an excessive temperature rise of the cooling water so that humidity control can be maintained.

Another gas direct-fired, absorption-cycle heat pump is being developed in a 36,000 Btu/hr (3-ton) nominal cooling capacity configuration by Columbia Gas System Service Corporation. The device utilizes a proprietary multicomponent, aqua-ammonia base, absorbent-refrigerant system and is undergoing performance testing and development in commercially available hardware.

CHAPTER 4

HEAT PUMP MANUFACTURING INDUSTRY

INTRODUCTION

There are about 50 heat pump manufacturers in the United States, depending on how they are enumerated among the various corporate relationships. About 70 are listed by the American Society of Heating, Refrigeration and Air Conditioning Engineers (ASHRAE), but there are many corporate ownership and subsidiary relationships among these, involving cooperative product line development, pricing and service. Further, such corporate relationships are always changing, so that it is difficult to provide an accurate picture of the entire industry at any one time. Table 4-1 lists the named manufacturers, where their headquarters are located, the capacity ranges of the system being offered, and the corporate relationships of most of the companies. Frequently, these extend through several levels, such as the ownership of Climate Master by Fredrich Air Conditioning which, in turn, is owned by Wylain, Inc., a relatively small conglomerate emphasizing capital equipment for residential and commercial buildings.

These heat pump manufacturers were surveyed for sales volume, number of employees and the types of heat pumps offered. The largest group (Table 4-2) was selected for direct query on their products and plans for near-term development. The major manufacturers, General Electric and Westinghouse, were added. They do not appear in the list because their heat pump activities represent too small a subdivision within these companies to appear separately in corporate public information releases required under U.S. securities law.

These companies may be further divided by a major distinction among their product lines. Most of the companies listed manufacture only air source units and these are usually restricted to air indoor coils. A representative group of their system for capacities below 60,000 Btu/hr

Table 4-1. Heat Pump Manufacturers

Manufacturers	Corporate Association	Headquarters	Capacity Ranges (000 Btu/hr)
AC Manufacturing		Cherry Hill, NJ	36 → 180
Amana Refrigeration		Amana, IA	<36, 36-180, > 180
Advance Design Assoc.		Miami, FL	<36, 36-180, > 180
American Air Filter		Louisville, KY	<36, 36-180
Addison	Distributors	Addison, MI	<36, 36-180
Armor Electric		Erie, PA	<36-180, 180
Bard		Bryan, OH	<36, 36-180
Barkow Manufacturing		Milwaukee, WI	<36
Bryant		Indianapolis, IN	<36, 36-180
Budco		Bloomfield, CT	<36, 36-180, > 180
Climate Master	Owned by Friedrichs of Wylain	Ft. Lauderdale, FL	<36, 36-180, > 180
Carrier A/C Division	United Technologies	Syracuse, NY	<36, 36-180, > 180
Century	Owned by Heat Controller	Jackson, MI	<36, 36-180 > 180
Command-Aire		Waco, TX	<36, 36-180
Day and Night	Division of Carrier	City of Industry, CA	<36, 36-180
Dunham-Bush		West Hartford, CT	<36, 36-180, > 180
Elmsbrook Refrigeration		Butler, WI	<36
Fasco		Fayetteville, NC	<36, 36-180
Fedders Corporation		Edison, NJ	<36, 36-180
Florida Heat Pump		Pompano Beach, FL	<36, 36-180
Friedrich A/C and Refrigeration	Owns Climate Master	San Antonio, TX	<36, 36-180
General Electric Central A/C Division		Louisville, KY	<36, 36-180, > 180
Heat Controller	Owns Century	Jackson, MI	<36, 36-180, > 180
Heat Exchangers, Inc.	Owns Koldware Division	Skokie, IL	<36, 36-180, > 180
Heil Quaker Corporation	Owned by Whirlpool	Nashville, TN	<36, 36-180
Henry Furnace	Owned by Luxaire	Elyria, OH	<36, 36-180
Infotran		New York, NY	36-180

International Heat & A/C	Owned by Wylain	Utica, NY	<36
Johnson Corporation	Owned by Magic Chef	Columbus, OH	<36, 36-180
Lennox		Marshalltown, IA	<36, 36-180
Luxaire	Subsidiary of Westinghouse	Elyria, OH	<36, 36-180
McGraw-Edison	Air Comfort Division	Albion, MI	<36, 36-180
McMillan Heat Pumps		Jacksonville, FL	36-180, > 180
Mammoth Division	Owned by Lear-Siegler	Minneapolis, MN	<36, 36-180, > 180
Marvair		Cordelle, GA	<36, 36-180
Modern Comfort		Fort Wayne, IN	<36
Mueller Climatrol		Piscataway, NJ	<36, 36-180
Patco	Pennsauken	Pennsauken, NJ	<36, 36-180
Singer	Climate Control Division	Carteret, NJ	<36, 36-180, > 180
Rheem Manufacturing A/C Division	Owned by City Investing Company	Fort Smith, AR	<36, 36-180
Royal A/C		Hialeah, FL	<36, 36-180
Ruud		Chicago, IL	<36, 36-180
Sun Dial	Owned by Square D Company	Mesquite, TX	<36
Supreme Aire		Santa Fe Springs, CA	36-180, > 180
Tappan A/C Division		Elyria, OH	36-180, > 180
Thermo Products		North Judson, IN	<36, 36-180
Vilter Manufacturing		Milwaukee, WI	180
Weatherking		Orlando, FL	<36, 36-180, > 180
Wesco Air Comfort		Beavertown, OR	<36, 36-180
Westinghouse Electric	Central Resid. A/C Division	Norman, OK	<36, 36-180
Whalen Company		Laurel, MD	<36
Whirlpool Heating and Cooling	Owns Heil Quaker	Nashville, TN	<36, 36-180
Williamson Company		Cincinnati, OH	<36
York Division	Owned by Borg-Warner	York, PA	<36, 36-180

Table 4-2. Major Heat Pump Vendors (excluding GE & Westinghouse)

Rank By Sales	Company	Employees	Sales ($ thousands)	Cumulative Sales
1	Rheem Manufacturing Co.	5000	550	550
2	Carrier A/C Division[a]	5500	(360)	910
3	York Division of Borg-Warner[b]	(5550)	(360)	1270
4	Fedders[c]	(4490)	294	1564
5	Amana Refrigeration, Inc.	4100	282	1846
6	Lennox	4000	(262)	2108
7	American Air Filter	5100	244	2352
8	Addison Products	1200	80	2432
9	Heil-Quaker	1200	65	2497
10	Fasco Industries, Inc.	2500	64	2561
11	Johnson Corp.[a]	800	(52)	2613
12	Williamson Co.	855	42	2655
13	Luxaire, Inc.	600	38	2693
14	Tappan A/C Div.[a]	533	(35)	2728
15	Vilter Manufacturing	420	22	2750
16	Heat Controller, Inc.	125	15	2765
17	AC Manufacturing Co.	200	13	2778
18	Board Manufacturing Co.[a]	150	(10)	2788
19	Command Aire Corp.	110	5	2793
20	Climate Master Products, Inc.	100	3	2796

[a]Sales inferred according to number of employees.
[b]Inferred according to 20% of Borg-Warner in 1976.
[c]Employees inferred according to amount of sales.

(15,000 kcal/hr) are reviewed below for major features and operating characteristics. A small second group of companies manufactures water-source heat pumps. The prominent characteristics of their product lines are also reviewed below.

AIR-TO-AIR SYSTEMS CURRENTLY BEING MARKETED

The firms listed here offer units with very similar operating characteristics and technical features. There are two standard outdoor temperatures used on test points: 47°F, for which the COP is usually between 2.4 and 2.9; and 17°F, for which the COP is usually less than 2.0. Technical features, such as the use of the cooling mode for defrosting and capillary tubes or thermostatic valves to help control refrigerant flow, are present in all of them. There are several outstanding features, however. Carrier has a line using two-speed compressors to modulate capacity, but there is no marked improvement in specific operating characteristics. Real value for this feature will probably show itself in maintenance and durability records. As noted in Chapter 3, Lennox has had a two-speed compressor under development for some time, but it is not yet being marketed. Chapter 5 mentions some of Texas Power and Light Company's favorable experience with the Lennox two-speed compressor. Another notable feature is the rotary compressor of Fedders which, again, has not led to a marked improvement in efficiency. Finally, it must be noted that York and Lennox have been using solid-state control modules. This is a very recent development with Lennox which, as recently as January 1978, relied solely on outdoor exchanger blow-through pressure differential to trigger the defrost cycle.

These distinctive features are noted along with high- and low-temperature heating capacities and COP in Table 4-3 for a set of 10 of the largest manufacturers of air-to-air equipment. By slim margins, Carrier low-speed operation achieves the highest high-temperature COP at 3.09 and York's 48,000 Btu/hr unit achieves the highest low-temperature COP at 2.00. The lowest high-temperature COP, at 2.3, is shown by the Fedders 48,000 Btu/hr unit and the lowest low-temperature COP at 1.57 by the Whirlpool 23,000 Btu/hr unit.

In connection with the specific provisions for defrost cycle initiation, the use of outdoor exchanger blow-through pressure differential (as used by the Lennox model described in this book) can fail rather disastrously under fairly common severe winter weather conditions. The unit in the table listed for Fedders uses a combination of timed cycle and outdoor coil temperature, so that it is not vulnerable. However, other Fedders units did use the pressure differential criterion and, in the abnormally severe winter of 1977-1978, many of the units experienced compressor breakdown. They were traced to erratic operation of the defrost cycle due to a combination of gusty winds and drifting snow over a long period. As a result the defrost mode was triggered sometimes unnecessarily and, what was worse, sometimes not triggered when ice had built up on the outside coil. Under such conditions, the compressor will work too hard under low lubrication conditions and eventually lock due to bearing friction buildup. The problem with the Fedders units became severe enough that persistent complaints to the Federal Trade Commission (FTC) eventually resulted in a consent order, agreed to by both the FTC and Fedders (*New York Times*, 23 February 1979; *Wall Street Journal*, 23 February 1979). Affected units had been manufactured between 1 November 1975 and 1 June 1978 and sold under brand names Fedders CICH and Climatrol. Repairs on up to 40,000 units were to be paid for by Fedders. All these units will now have their defrost cycle triggering revamped to avoid sole reliance on pressure differential.

Table 4-3. Operating Features of Air-Source Heat Pumps (capacity in units of 1000 Btu/hr)

Manufacturer	Model (indoor coil)	Capacity	COP	Capacity	COP	Distinctive Features
Amana	EPH03600-1A	38.0	2.80	17.0	1.90	
Amana	EPH02400-1A	25.0	2.90	13.0	1.70	
Carrier	T2042 Hi speed	46.0	2.87	26.0	1.95	Two-speed
Carrier	T2042 Lo speed	29.5	3.09			Compressor
Fasco	H710536	38.0	2.60	22.5	1.80	Thermostatic
Fasco	H710530	31.0	2.40	15.0	1.40	Expansion valve
Fedders	CKA048COA	47.0	2.30	28.0	1.60	Rotary compressor
Fedders	CKA024COA	23.5	2.60	13.0	1.80	
General Electric	WE120C	120.0	2.94			
Lennox	CHP10-651	61.0	2.5	34.5	1.70	Solid-state control module
Lennox	CHP10-411	38.0	2.4	21.5	1.70	
Rheem	RCPB-A048	49.0	2.60	28.5	1.90	
Rheem	RCPB-A018	18.0	2.50	11.0	1.70	
Westinghouse	HP060D	63.0	2.84			Subcooling criterion for
Westinghouse	HP048D	55.0	2.90			refrigerant control
Whirlpool	NPHH047	49.0	2.52	30.0	1.76	
Whirlpool	NPHH023	24.0	2.43	14.0	1.57	
York	CHPA48	49.0	2.90	26.5	2.00	Solid-state control module
York	CHPA24	23.5	2.8	13.0	1.95	

Table 4-4. Features of Water-Source Heat Pumps (capacity in units of 1000 Btu/hr)

			Entering Water Temperature				
			60°F		80°F		
Manufacturer	Model	Flowrate (gal/min)	Capacity	COP	Capacity	COP	Distinctive Features
Climate Master of Friedrichs	52	7	51.0	2.63	65.0	2.91	Precision sizing of capillary tubes
	22	3	21.5	2.43	28.4	2.67	
Weatherking	54-1	14	54	2.80			Provision for accessory water flow control
	34-1	9	34	2.80			
Mammoth of Lear Siegler	72	10	72	2.80			

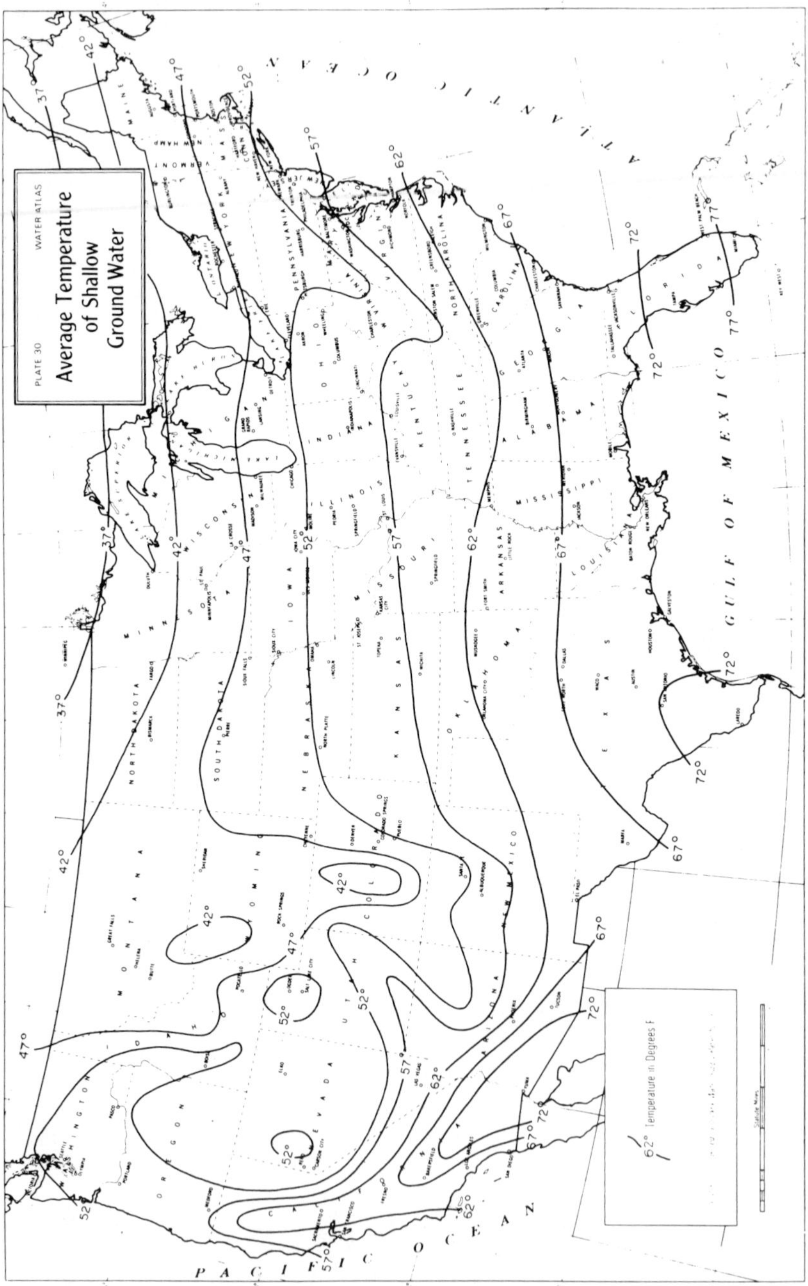

Figure 4-1. Average temperature of shallow groundwater in the U.S.

WATER-TO-AIR HEAT PUMPS

Even more than with air-to-air heat pumps, water-to-air products have largely similar operating characteristics and technical features. A representative set of three manufacturers (very few manufacturers make water-source heat pumps) is reviewed in Table 4-4. The units are all relatively simple in their control features because the operational conditions are presumed to be constant.

It is important to realize that their COP of around 2.8 will be maintained against drastic fall in the outside temperature. However, they all need a water reservoir at 60°F or better which would cover most of the heavily populated temperate and northern areas of the U.S. (Figure 4-1). Some utilities have been experimenting with water-source heat pump operation (see Chapter 5) below 60°F, but they use special models and have experienced some difficulties. As noted in Chapter 3, Climate Master is developing a line that can effectively use water down to 45°F. It also expects considerably improved COP. This will open its market to the northern states where the lower temperature water tables will allow better cooling performance.

CHAPTER 5

HEAT PUMP DEVELOPMENT PROGRAMS AND MARKET PROSPECTS

INTRODUCTION

Heat pumps are being promoted through many government and utility programs. The federal government is interested in its potential to reduce oil import requirements by:

- shifting from a gas-oil combination for home heating to electricity (at comparable total energy efficiency), which may be generated without those fuels;
- more effective use of gas in gas-fired heat pumps; and
- better load balancing of utility demand for a reduced generating requirement, allowing less use of oil-fired and older inefficient plants.

State and local governments are interested in electric heat pumps to alleviate economic hardship in rural residential heating where only electricity may be available as a distributed energy resource. Electric utilities are interested in load balancing capability and a way to dissuade electric-heating customers from switching to gas or oil as electricity rates rise. Gas supply utilities have been promoting heat pump developments, as described in Chapter 3, to counter the movement toward electric heat pumps. The various programs of these groups and an outlook for the heat pump market are described in the following sections. Utility programs aimed at the most immediate gains are presented in the first section. Government programs in technical development and market promotion through subsidy are the subjects of the next two sections. The last section discusses estimated market trends for the near term and limitations that may be imposed on them.

UTILITY HEAT PUMP PROGRAMS

Of the several hundred operating electric utilities in the U.S., those promoting heat pump use or development comprise a minority. A representative group, compiled by the Electric Power Research Institute (EPRI), shows many utilities that have programs directed specifically to their particular situations. The underlying circumstances that distinguish their heat pump requirements involve the local weather, relative costs and availability of fuel and electricity, and special government tax benefits. Heat pump-related activities currently engaged in by 24 utilities and utility associations are discussed here. Table 5-1 provides an overview of the scope and diversity of their promotional and research activities to indicate how these may be related primarily to the relevant features of climate in their service area.

There is a sharp orientation toward the special interests of the utility, along with a fair regard for the special needs of their customers. Outside of DOE sponsorship of these programs, it appears that utility programs are being pursued with greater faith that present difficulties can be overcome than shown in purely governmental programs. The problem in the most heavily populated areas of the U.S. (where the important market would lie) is a winter with very low nighttime minima. Utilities located in these areas are clearly aiming their efforts at alleviating this specific difficulty through:

- energy-conservative home design to reduce total heat pump requirements;
- thermal storage to modulate duty cycle and possibly allow shifts that tend to balance total utility annual or diurnal load variation;
- solar assistance for thermal storage; and
- use of groundwater source as annual thermal storage.

Outside of the severe northern climate condition that still works against a heat pump market, there are two other typical U.S. conditions. In the south, central and east, with a combination of mild winters and warm summers, there is already a widespread acceptance of heat pumps, as evidenced by the utility programs of Alabama Power, Gulf States and Texas Power and Light. Confidence has been established for their service areas, and the utilities see some return in sponsoring heat pump usage and development. In the Northwest, with the gentlest of U.S. climates, heat pumps are slowly being accepted as a way to avoid departure from electricity as the sole energy service. The U.S. Southwest, barely represented in the table, clearly will be preoccupied in utilizing solar energy by virtue of special California tax law benefits added to those of the new federal tax law amendments.

Table 5-1. Utility Heat Pump Programs

Utility or Association	Relevant Climate Features	Heat Pump Use Promotion	Experiments and Demonstration
Alabama Power	Mild winter, warm summer	Comprehensive service and recordkeeping	Small "good sense home" conservation
American Electric Power	All of U.S.	None	Thermal storage for load balance
American Public Power Association	All of U.S.	Lobby for federal and state subsidy	None
Arkansas Power and Light	Mild winter, warm summer	Individual customer persuasion	Collaborate with general EPRI monitoring
Association of Edison Illuminating	All of U.S.	None	Cosponsored with EPRI (Blake and Gernet, 1978)
Central Hudson Gas & Electric	Low-winter minima, cool summer	None	Demonstration project on energy-efficient residence with solar
Environmental Edison	Low-winter minima, warm summer	Issuance of standards and certification	None
Consolidated Edison of New York	Cool winter, warm summer	None	Demonstration in new apartment projects
Easton Utilities Commission	Mild winter, warm summer	Considering if tests prove favorable	Test of water source on city water in air
Edison Electric Institute	All of U.S.	None	None
Gulf States Utilities Company	Mild winter, hot summer	Inspection of installation	Demonstration projects on "energy-sensible home"
National Rural Electric Coop Association	All of U.S.	None	Research on heat pump for hot water
Niagara Mohawk Power	Cold winter, cool summer	None	Evaluation experiments for northern climates
Northern States Power Company	Cold winter, cool summar	Information dissemination	Demonstration on energy-conservative home, "Roseville"

Table 5-1, continued

Utility or Association	Relevant Climate Features	Heat Pump Use Promotion	Experiments and Demonstration
Pacific Power and Light Company	Mild winter, cool summer	Recommend for new construction	Demonstration on "TERA ONE" energy-conservative home
Pennsylvania Power and Light Company	Cool winter, warm summer	None	Experiments emphasizing reliability
Portland General Electric Company	Mild winter, cool summer	None	Experiment with solar-assisted heat pumps
Potomac Electric Power Company	Mild winter, warm summer	Information dissemination	Sponsor experiments with solar-assisted heat pumps and thermal storage
Public Service of New Mexico	Warm winter, hot summer	None	Collaborated with EPRI (Blake and Gernet, 1978)
Public Service Electric and Gas Company	Cool winter, warm summer	Information dissemination	Cosponsoring thermal storage studies
Southern California Edison Company	Mild winter, cool summer	None	Experiments with solar-assisted heat pumps
Texas Power and Light Company	Mild winter, hot summer	Past service program—no longer necessary	Experiments with solar-assisted Lennox heat pump systems
Tennessee Valley Authority	Mild winter, warm summer	New heat pump service program	Small solar-assisted heat pump demonstration
Electric Power Research Institute	All of U.S.	None	Widespread coordination and cosponsoring R&D

Utility heat pump programs are described in further detail later in this chapter.

Alabama Power Company (APC)

By far the most comprehensive heat pump promotion program is being run by the Alabama Power Co. which, in 1967, started a service policy to customers having heat pumps installed known as the Assured Heat Pump Service Program. Units that had been installed as early as 1964 were taken into the program to be serviced with full recording of all service requirements until they completed 10 years of service. At present, there are 9100 units covered in the program, all in residential service except for about 500 serving commercial or industrial buildings. There have been as many as 15,000 units in the program. To be eligible, the heat pump must be supplied by a manufacturer on an approved list. This included 16 manufacturers at the program's start; currently, however, there are only 9. About 5 of these manufacturers have supplied between 75 and 80% of the units now on the program.

The service offered is a 10-year contract for service and repair of the heat pump, either directly by APC or by a few dealer-service agencies in the APC area. The utility supplies power to many customers in relatively isolated rural locations where local dealers cannot assure service. The APC-supplied service is especially important to extend the option of heat pump usage to these customers. Running charges for the contract are set at the time of installation and remain unchanged for the life of the contract. Initially, they were $1/ton/month, with a 3-ton minimum. By 1975, new contracts cost $2/ton/month. One of the most important features of the program is the detailed, uniformly formatted service record maintained by APC.

The purpose of the Assured Service Program is primarily to support heat pump sales and use in Alabama, anticipating that this would build good will among customers and then help balance the load placed on APC's grid. The quality of heat pump dealers would be developed through having to meet minimum qualifications for participation. In addition, the following desired effects would be realized:

1. Customers would be assured of good equipment and adequate service for a reasonable charge.
2. Dealers would receive benefits as a certified dealer using standards uniformly set and authenticated by the power company.
3. Wholesale distributors would gain through increased sales and improved installations from dealers.

4. Manufacturers would benefit by having equipment installed using better than normal installation practices and by having available, periodically, service histories on all units in Alabama covered under the ASP.

The manufacturers currently in the program, according to Lovvorn (1975), are:

Carrier	Fedders
General Electric	Amana
Westinghouse	Lennox
Bard	Singer
Day & Night	

Table 5-2 displays the defect frequency for a number of major defect categories, for the units of each manufacturer in the program (Orange, 1977). As part of the program's arrangements, the specific manufacturer for each row of figures is not identified. It is quite clear that compressor, refrigerant and power problems predominate with wide variation among the manufacturers in susceptibility to each type of problem.

In the service program, each unit covered is assigned a number which identifies it for all services. These are recorded individually for each unit with characteristics of the service call according to defect and remedial action, indicated in accordance with the code of Table 5-3.

In the various compilations of service requirements, it is clear that the research provides a data base for comparing categories of units over a time period comparable to their useful life. They are distributed 60-40 among split and unitary systems. Supplemental energy must be derived from resistance heating. Currently, there are no Carrier two-speed or Fedders rotary compressor units in the program; 75-80% of heat pumps are supplied by five manufacturers.

Typical weather conditions for Birmingham heat pump usage are somewhat milder than those for the U.S. as a whole. However, since 1976, Birmingham has experienced some long low-temperature periods, and there have always been high humidity levels to trigger defrost cycles. Some units experienced serious problems with solid-state devices sensing the defrost requirement. Moisture entered the devices to create inaccuracies in their sensing of refrigerant and outside air temperatures.

Among experiments in conservation, APC is testing heat pump units in a well-insulated "good sense" home. They expect differences in long-term operational effects because of the difference in balance between energy required for heating and cooling. Normally, the heating load near Birmingham is 1.5 times the cooling load. With the "good sense" home's better insulation reducing the heating requirement, the loads are now

Table 5-2. Alabama Power Company Assured Heat Pump Service Defect Frequency Analysis Comparison on all Units Currently Under Contract as of September 30, 1975 (Manufacturers with 50 or more units under contract) (Orange, 1977)

					Compressor Problems				Refrigerant				Fan Problems															
			100-199		400-499				200-299		300-399		500-529		823		530-539		540-599		600-629		630-699		700-729		730-751	
	Number of	Total Years Under	Internal		Motor Circuit		Failures[a]		Leaks		Components		General		Belts[b]		Defrost Circuit		Transformers Unit Housing Drain Pan		Thermostats Indoor & Outdoor		Supple- mental Heaters		Ref. & Oil Charge Adjustment		General Maintenance Dirty Coil, etc.	
Mfg.	Units	Contract	D	DR	D	DR	F	FR	D	DR	D	DR	D	DR	D	DR	D	DR	D	DR	D	DR	D	DR	D	DR	D	DR
				7.1		14.0		7.1		2.7		5.2		14.0		3.4		4.6		2.5		1.2		5.2		0.4		0.2
				3.8		25.5		3.4		6.8		3.7		10.6		13.7		6.1		1.0		2.2		6.1		0.2		0.1
				2.6		13.0		2.6		5.5		2.0		19.8		0.2		6.8		0.2		1.8		7.0		0.0		0.9
				3.7		5.0		3.7		5.5		7.1		11.9		5.1		5.5		1.3		2.5		3.1		0.4		0.2
				13.4		29.1		13.8		20.5		26.2		25.3		1.4		4.8		0.5		1.0		21.4		2.4		0.0
				5.1		12.3		5.1		8.7		13.7		17.9		3.8		11.6		1.3		2.3		5.2		0.8		0.2
				6.9		11.8		7.3		11.0		2.7		25.6		11.8		2.8		0.7		2.7		13.2		0.2		0.5
				9.3		18.8		9.3		7.0		16.4		21.8		6.0		7.8		1.0		3.7		8.8		0.3		0.1
				14.9		13.8		14.7		11.4		6.2		16.5		5.7		12.1		2.3		2.8		11.7		0.2		0.7
				7.9		24.5		7.9		7.5		5.6		28.7		0.0		15.8		1.6		2.6		5.0		0.9		0.4
				7.6		8.2		7.5		16.9		5.6		15.7		5.0		5.1		2.9		3.1		5.7		0.8		0.2
All Others	62	296.1	34	11.5	62	20.9	34	11.5	71	24.0	63	21.3	97	32.8	41	13.8	18	6.1	7	2.4	12	4.0	31	10.5	10	3.4	2	0.7
Totals	8,521	31,832.1		6.0		15.5		5.9		8.1		8.0		19.0		4.4		9.9		1.5		2.5		5.3		0.8		0.2

Legend:

D = Actual Number of Defects

$$DR = \text{Defect Rate} = \frac{\text{Actual Number of Defects}}{\text{Total Years Under Contract}} \times 100$$

$$FR = \text{Failure Rate} = \frac{\text{Total Compressor Failures}}{\text{Total Years Under Contract}} \times 100$$

[a]Failures are defects which are also included in the 100-199 and/or 296 code groups.

[b]Fan belts are not included in 500-529 General Fan Problem group.

Table 5-3. Alabama Power Company Electric Heat Pump Service Call Code Card (Alabama Power Co.)

I. Reason for Service Call

A — No Heat
B – No Cooling
C – Insufficient Heating
D – Insufficient Cooling
E – Excess Humidity
F – House Drafty
G – House Stuffy
H – Uneven Temperature
J – Indoor Noise
K – Outdoor Noise
M – Condensate Problem
N – High Electric Bill
P – Service Inspection
Q – Unit Start Up
R – Call Back
S – Other (Explain)

II. Servicer's Analysis

A – Power Supply
B – Protective Device Open
C – Improper Refrigerant or Oil Charge
D – Electrical Circuit or Component Defect
E – Refrigerant Circuit or Component Defect
F – Unit Noisy
G – Hot Gas Pulsation
H – Improper Air Flow
J – Improper Air Distribution
K – Improper Capacity
M – Enclosure or Insulation Defect
N – Condensate Drip Problem
P – Defrost Problem
Q – Control Circuit Defects
R – Accessory Defect
S – Installation Problem or Check Out
T – Leak (Refrigerant)
U – Other (Explain)

III. Corrective Action

A – Replace Unit
B – Replace Cycle
C – Replace Compressor
D – Install Pack
E – Install Kit

Corrective Action, cont.

F – I.D. Field Repair or Adj.
G – O.D. Field Repair or Adj.
H – I.D. Component Repl.
J – O.D. Component Repl.
N – Installation Corrections or Check Out
P – Reset Switch
Q – Customer Education
R – Other (Explain)

IV. Defect Code

Compressor Problems (100)

110 Compressor Acid
120 Compressor Defects (Mechanical)
 1 Won't Run
 2 Pulls Locked Rotor Amps
 3 Won't Pump
 4 Leak Back
 5 Poor Pumping
 6 Pulls High Amps
130 Compressor Defects (Electrical)
 1 Windings Open
 2 Windings Shorted
 3 Windings Grounded
140 Compressor Noisy
 1 Running Noise
 2 Starting-Stop Noise
 3 Vibration
150 Compressor Motor Internal Winding Protector Defects
 1 Protector Open
100 Compressor (Other)
 1 Wrong Size, Type, or Phase
 2 Terminal Defects

Refrigerant Leaks (200)

200 Coil
 1 Saddle Joint
 2 Brazed Joint
 3 Mechanical Joint
 4 Chafing
 5 Split Tube
 6 Broken Tube
210 Factory Tubing
 1 Saddle Joint
 2 Brazed Joint
 3 Mechanical Joint

Table 5-3, continued

Refrigerant Leaks, cont.

Code	Item
214	Chafing
5	Split Tube
6	Broken Tube
220	Field Tubing
1	Brazed Joint
2	Mechanical Joint
3	Chafing
4	Split Tube
5	Broken Tube
230	High Pressure or Defrost Term. Control
1	Brazed Joint
2	Mechanical Joint
3	Broken Tube
4	Bellows
235	Low Pressure Control
6	Brazed Joint
7	Mechanical Joint
8	Broken Tube
9	Bellows
240	Metering Device (Exp. Valve, Cap. Tube, Etc.)
1	Brazed Joint
2	Mechanical Joint
245	Liquid Line Service Valve
6	Brazed Joint
7	Mechanical Joint
8	Schrader Valve
250	Vapor Line Service Valve
1	Brazed Joint
2	Mechanical Joint
3	Schrader Valve
255	Pilot Valve (for Reversing Valve)
6	Brazed Joint
7	Mechanical Joint
260	Filter–Drier
1	Brazed Joint
2	Mechanical Joint
265	Sight Glass
6	Brazed Joint
7	Mechanical Joint
270	Check Valve
1	Brazed Joint
2	Mechanical Joint
275	Reversing Valve
6	Brazed Joint
7	Mechanical Joint
280	Fusible Plug

Refrigerant Leaks, cont.

Code	Item
281	Mechanical Joint
2	Brazed Joint
285	Muffler
6	Brazed Joint
7	Mechanical Joint
290	Modulator/Accumulator
1	Brazed Joint
2	Mechanical Joint
295	Other (Explain)
6	Compressor Welds
7	Compressor Service Couplings
8	Compressor Hermetic Service Fittings

Refrigerant Components (300)

Code	Item
310	Metering Device (Exp. Valve, Cap. Tube, Etc.)
1	Flooding
2	Starving
3	Erratic
4	Restricted
5	Bulb Loose
6	Power Assembly
320	Coil
1	Restricted
2	Wrong Circuit
3	Leak Unrepairable
4	Fins Loose
5	Scale or Corrosion
330	High Pressure or Defrost Term. Control
1	Electrical Parts
2	Calibration High
3	Calibration Low
340	Low Pressure Control
1	Electrical Parts
2	Calibration High
3	Calibration Low
350	Filter-Drier
1	Plugged
2	Installed Backwards
3	Core
360	Check Valve
1	Stuck
2	Installed Backwards
3	Noisy
370	Reversing Valve
1	Internal Leak Back

Table 5-3, continued

Refrigerant Components, cont.

Code	Item
372	Stuck
3	Chatters
4	Pilot Valve Stuck
5	Coil
300	Components (Other)
1	Modulator/Accumulator
2	Muffler
3	Service Valve
4	Quick Connects
5	Sight Glass/Moisture Indicator
6	Schrader Valve
7	Strainer

Compressor Motor Circuit Components (400)

Code	Item
410	Compressor Wiring
1	Wired Wrong
2	Shorted
3	Grounded
4	Broken
5	Fuse-Winding Protector
420	Motor Starter/Contactor
1	Stays Open
2	Stays Closed
430	Start Capacitor
1	Open
2	Shorted
3	Blown
440	Run Capacitor
1	Open
2	Shorted
3	Blown
450	Start Capacitor Relay
1	Coil Open
2	Coil Grounded
3	Contacts Stuck or Burned
460	External Overload in Line–No. 1
1	Stays Open
2	Internal Short
3	Grounded
470	External Overload in Line–No. 2
1	Stays Open
2	Internal Short
3	Grounded
480	Discharge Line Thermostat
1	Stays Open
2	Shorted
3	Grounded

Compressor Motor Circuit Components, cont.

Code	Item
490	Cir. Breaker (Factory Supplied)
1	Won't Reset
400	Compressor Circuit (Other)
1	Motor Starter Aux. Relay
2	Time Delay Relay
3	Crankcase Heater
4	C.H. Relay

Other Components (500)

Code	Item
510	Fan Motor
1	Noisy
2	Vibration
3	Capacitor
4	Relay
5	Overload
6	Pully
520	Fan Blade or Blower Wheel
1	Noisy
2	Vibrates
3	Belt
4	Pulley
5	Bearings
6	Speed
530	Defrost Circuit
1	Initiation Device
2	Adjust Initiation or Termination
3	Termination Device
4	Air Hose/Tube
5	Relay
540	Transformer
1	Grounded
2	Primary Open
3	Secondary Open
4	Noisy
5	Fuse-Control Circuit
550	Sheet Metal (Unit)
1	Rusted
2	Finishing Flaking or Peeling
3	Damage (Dented, Scratched, etc.)
4	Insulation Loose
560	Drain Pan
1	Leaks
2	Rusted
3	Freeze Up
500	Component (Other)
1	Auxiliary Relay
2	Other Relay

Table 5-3, continued

Code	Description
Accessories (600)	
610	Indoor Thermostat
1	Out of Calibration
2	Not Level
3	Wired Wrong
4	Improperly Located
5	Loose or Broken Wire
620	Outdoor Thermostat
1	Out of Calibration
2	Improperly Located
630	Supplemental Heater
1	Contactor/Relay
2	Thermal Cutout
3	Wiring
4	Fusible Link/Manual Reset
5	Element
6	Manual Heat Switch
7	M.H.S. Relay
640	Low I.D. Air Flow Indicator
1	Complete Indicator
2	Miscellaneous Parts
600	Accessory (Other)
1	Grille
2	Diffuser
Misc. Service (700)	
710	Refrigerant Charge Adjustment
1	Overcharge
2	Undercharge
720	Oil Charge Adjustment
1	Overcharge
2	Undercharge
3	Clean Up
730	Compressor Valve Plate
1	Complete Set (One Valve)
2	Miscellaneous Parts
750	Maintenance (General)
1	Coil Dirty
2	Preventive Maintenance
3	Factory Field Fix
770	Found No Defect
780	Incomplete Call
1	Checked for Difficulty Only
Installation & Application (800)	
810	Unit Selection
1	Under Sized
2	Over Sized
Installation & Application, cont.	
813	Location Incorrect
4	Misapplied
820	Contributing Conditions
1	Building Defects
2	Weather Extremes
3	Improper Insulation
4	Other (Explain)
830	Duct System
1	Improper Air Flow
2	Unbalanced Air System
3	No Filters
4	Air Leakage
5	Improperly Insulated
840	Power Supply
1	Under Sized
2	Wiring Error
3	Circuit Breaker
4	Fuse
5	Utility Problem
850	Control Wiring Error (Field)
1	Damaged
2	Control Wiring Error (Factory)
860	Plumbing or Tubing Error
1	Condensate Drain
2	Refrig. Tubing Incorrect Size
3	Refrig. Tubing Uninsulated
800	Installation & Application (Other)
1	Original Start Up
Customer Education (900)	
910	Electrical Controls
1	Indoor Thermostat
2	Fuse Blown/Breaker Tripped
3	Fuse Loose
920	Duct System
1	Filter Dirty
2	Diffuser Adjustment
930	Comfort General
1	Comfort Humidity
2	Defrost
940	Furniture Location
1	Furniture Affecting Thermostat
2	Furniture Affecting Distribution
950	Equipment Operation
1	Cost of Operation
900	Customer Education (Other)

evenly balanced. APC is presently building an experimental home using a solar collector for supplemental heat supplied to a low-temperature rock storage unit. It will assist the heat pump, but not necessarily to avoid low temperature, low COP conditions. In another experiment, APC is using a heat pump to provide hot water and is testing now to see whether there are any new problems introduced in this application. There has been DOE sponsorship of a project to develop a water-heating heat pump, which would cost $325-375. The compressor for the unit would cost $100-150. Comparing such costs to $100-125 for an electric hot water heater and the APCF rate of 5¢/kWh leads to a one-year payback period.

The APC eventually would like to relinquish its program but only under conditions that would assure that customers would be left in good hands. These conditions include manufacturers meeting APC criteria on heat pump operating characteristics and offering their own 10-year warranty. They regard a 5-year term as little more than a compressor warranty. Many current service contracts have limitations on cost totals and per service calls. The utility feels that such a policy would give the manufacturer a better idea of the operating characteristics of their heat pumps under actual extended service conditions. Reliability will be a serious issue in the near future, especially as many home builders will be seeking the cheapest possible heat pumps to meet using demand. The 10-year warranty will compel technical attention to reliability problems.

American Electric Power (AEP)

American Electric Power is a holding company which, through its subsidiary American Electric Power Service Corporation, provides services such as marketing, research and development to associated small utility companies. It has been particularly interested in the use of thermal storage systems associated with heat pumps as electric resistance heaters. Experiments were carried out both for low-temperature liquid and high-temperature hot-rock concepts and were aimed at moderating the diurnal demand variation. AEP found that using a large water vessel as a supplement to the heat pump was not a good economic investment because there was insufficient improvement in reasonably averaged COP. However, the hot-rock type can be charged by resistance heating at night to 1400°F. As such, it can be projected as economically sound when time-of-day differential electric rates are applied. The payback period for the extra investment of $650 is less than 5 years. Of course, its use is associated with resistance heating. The storage unit being tried is a ceramic furnace available from the TPI Corporation of Johnson City, Tennessee. AEP has not yet decided whether to recommend the system for all its customers. At present, the best response to the home heating problem by electricity appears to be heavy thermal insulation of the home with off-peak charging by resistance heating of the thermal storage unit and circulating air flow most of the time.

American Public Power Association (APPA)

About 1400 publicly owned utilities comprise the American Public Power Association. Programs to promote heat pump usage via demonstration experiments, information and service arrangements are all done under the aegis of the separate utilities. No formal coordination of their efforts is undertaken by APPA, which primarily recommends heat pumps as a conservation measure to replace electric resistance units. It also lobbies for state government promotional benefits in regions where it thinks the heat pump is applicable. The association is cooperating with EPRI in some data gathering programs.

Arkansas Power and Light Company (APL)

The Arkansas Power and Light Company serves about 40,000 residential customers who are totally dependent on electricity for space temperature conditioning. They are located in rural areas where there is no gas available. Between 10,000 and 12,000 heat pumps are already installed. The utility has no formal program to foster heat pump use. Their principal method is a one-on-one marketing approach with their customers.

Under closely controlled conditions, APL has carried out experimental studies of the efficiency of heat pumps for the particular requirements of its own customers. APL does not disclose the funding for these studies but does release some particulars of collaborating organizations. Associated with the general EPRI heat pump study program during the year 1976, it installed load survey recording devices on 10 heat pump loads in the Warren, Arkansas, area and one "Energy-Saving Home" in the Stuttgart, Arkansas, area. It was able to correlate the effects of weather changes on heat pump performance. For 1977, it contracted with the University of Arkansas to examine total energy consumption and shifts and moderations of peak demand for the "Energy-Saving Home." Because of its high thermal insulation, it experienced the common paradox that the heat pump necessarily spends a greater proportion of its operating time during cold weather conditions and, therefore, manifests reduced seasonal coefficient of performance, even though total seasonal operating costs are lower. A third project, centering on heat pumps with thermal storage systems, is ongoing in-house and is scheduled to run to 1 February, 1981. APL is studying 12 residential and 7 commercial applications of hot and cool thermal storage systems associated with solar heating and heat pumps. These systems are monitored to determine the most effective equipment/storage system on APL's system load, observe design parameters, and acquire other information that will be of use in advising its customers. In general, it has found thermal storage by itself not economical. It was thought that the system would enable the customer to build up a reserve by operating his heat pump at off-peak time; however, it has been found that there is some benefit to using a water tank storage attached to a solar supplement. Hot thermal storage systems were made by General Electric and Carrier. They are also doing some work on cool storage with potential DOE sponsorship because APL also has a summer peak load that they would like to reduce. In this work, they are using as a prototype water storage unit, an A.O. Smith 120-gallon electric water heater surrounded by 2½ inches of polyurethane insulation. They use one tank for each ton (12,000 Btu/hr) of house heating or cooling load. The tank is modified by installing 3/8 inch copper tubing around which, in typical application, a 1-inch-thick annular layer of ice forms. By actual test, it was found that the tank could withstand a total freezeup for a total cooling load of 11 ton-hours (132,000 Btu, 33,300 kcal). In normal operation for both heating and cooling, the house draws on the storage unit; only when storage is exhausted is the heat pump used directly.

Association of Edison Illuminating Companies

This organization is an association of 91 investor-owned electric utilities, which sponsor a variety of studies on issues of general interest to the member utilities. It recently joined with the Electric Power Research Institute to sponsor a study by Westinghouse (Blake and Gernert, 1978) for obtaining information on residential heat pump heating system operations based on field tests of existing installations (see Chapter 1). Heat pump and total house electric energy usage were recorded every 15 minutes. A total of 120 existing installations were instrumented in 12 different geographic regions. The Association considered the following conclusions of the study most relevant to utility operations:

1. Average SPF could be related to degree days heating.
2. About 15% of the calculated SPFs were within ±10% of a predicted ideal SPF; 25% were within ±15% of the predicted ideal SPF.
3. Heat pump heating systems should save considerable energy relative to electric furnaces over the wide range of degree days heating encompassed by this test.
4. Group heat pump system electrical demand and group total electrical demand peaked in the afternoon during cooling months and in the morning during heating months.

5. Group heat pump system demands share a higher coincidence factor than group total house demand because of the outdoor temperature as a common factor influencing heat pump demand.
6. Not counting periodic maintenance calls, there were approximately 1.3 service calls per heat pump year with those requiring service averaging about 2 repairs each.

Central Hudson Gas and Electric Corporation (CHGE)

This utility operates in upper Westchester and north neighboring counties in New York state, providing both gas and electricity. It is unusual, as a gas supplier, to be conducting studies on electric heat pump usage. These include: (1) system incorporation of thermal storage; (2) heat pumps for commercial buildings; (3) the demonstration project "Wenning House," and (4) a study of heat pump usage with oil heat supplement. All these projects are being carried out at annual costs of $30,000. The "Wenning House" is an energy-efficient residence built by architect Harry Wenning. The project was designed to determine relative energy inputs from the solar, heat pump and electric resistance heating components during different weather patterns on a seasonal basis. The solar system component is very large, and energy needs for the home have been met primarily through it. In three heating seasons the heat pump has yet to require maintenance, although this could well be due to the low usage. For the study of oil backup with heat pumps, much smaller houses (value approximately $50,000) subject to more realistic conditions are being used. Its goal is to develop quantitative criteria on the impact of heat pump systems in conjunction with stored energy in the form of oil on the utility load, and to establish the effectiveness of this system for energy conservation. Using a solar supplement in conjunction with a heat pump has been found to be a risky investment. For example, it is impossible to justify a $10,000 solar supplement to work with a $3,000 heat pump, when the major energy savings are already realized by the heat pump.

Commonwealth Edison Company (CE)

This utility supplies electricity to metropolitan Chicago. Currently it is promoting heat pumps to ease its electrical load problems. The program assures realiability of heat pump installation and a minimum level of performance in the Chicago area.

In particular, CE is concerned that the SPF is usually determined on the basis of estimated load. The limited metered test data developed to date reflect the uncertainty of comparing data from two different locations and/or years. Therefore, CE is participating in a joint field test program with cooperating manufacturers to establish representative SPFs for their heat pumps in the Chicago market. It hopes the test data will indicate factors that can be used to modify theoretical SPFs to values that can be used to realistically forecast energy consumption. In the initial phase, the manufacturer provides two complete heat pump systems designed to facilitate operation as a heat pump and as an electric furnace. CE will install the two heat pumps and the associated test metering, and take readings as required. The test programs will extend over at least one complete heating season. The test data for the heat pumps will be made available to the participating manufacturers.

CE has designed a special metering package that will automatically record every midnight the energy consumption during the previous 24-hr period, then alternately switch the unit from heat pump to electric furnace operation. Such a procedure will facilitate establishment of a credible SPF. To qualify for this program, a manufacturer's heat pump must demonstrate an SPF of at least 1.5.

CE does not intend to sell or service the endorsed heat pumps. All work will be solely in the hands of the contractor; however, the utility will support those brands that satisfy the SPF and other requirements, as outlined in the following extract from the brochure they have prepared:

"Commonwealth Edison Company will endorse and lend its support to the sale of heat pumps. This endorsement and support will be limited to those brands that have satisfied our requirements for participation. These are:

1. Certification by the American Refrigeration Institute.
2. Must achieve an actual installed seasonal performance factor of at least 1.5:1, compared to an electric forced air furnace in the same home. This performance factor is determined over a period of one complete heating season as part of Edison's field evaluation program.
3. Must operate as reliably as a flame type forced air furnace in combination with an electric air conditioner.
4. Must make available, for purchase by the customer, an extended service contract through the first five years of operation, on a competitive basis with flame type forced air furnaces and electric air conditioners.
5. Must be installed by dealers who:
    a. have attended substantive training sessions on installation and service of heat pumps sponsored by the manufacturer, and are certified by the manufacturer to install and service their heat pump equipment.
    b. will provide the customer with prompt and reliable service, and recommend the customer purchase a service contract.
6. Must have parts stocked in the immediate area.

7. Must be supplied with a convenient means by which the supplemental resistance heaters can be controlled by the indoor thermostat, should a problem develop in the refrigeration cycle.
8. As an integral part of the installation, should be equipped with sufficient electric resistance heating to maintain indoor temperature at outdoor design conditions, without the operation of the refrigeration cycle.
9. Must have outdoor thermostats connected to each major increment of supplemental heaters, above the initial increment staged to operate at successively lower temperature settings in relation to the heat pump balance point.
10. The distributor or manufacturer should provide assurance to the customer that the equipment has been properly sized and installed.
11. The manufacturer/distributor must supply the customer with a recourse source should the contractor not provide proper and prompt service."

Consolidated Edison of New York

Of all U.S. utilities, this probably serves the largest population and population density over its service region. Its charges, at 9-10¢/kWh, are about the highest in the U.S. and are due to a combination of high costs of fuel and power plant construction in the nation's most heavily developed area. Support of customer heat pump use is primarily in participation in heat pump use experimental projects. The utility is one of the co-sponsors with EPRI of the Westinghouse study (Blake and Gernert, 1978) on load and use characteristics in 120 heat pump installations. In addition, it has experimented extensively with its own generator exhaust as a source for central heat pumps to heat a housing project.

The study, funded by DOE, determined the feasibility of heating and cooling North Town Phase II of Roosevelt Island housing development with central heat pumps. Roosevelt Island is one of several large islands in the East River separating Manhattan from Brooklyn and Queens. For heating, the discharge water from CE's Ravenswood Station on the East River opposite the housing site was used; in summer, East River water was used as a heat sink for cooling. The heat pump scheme was compared with two conventional systems: (1) baseboard resistance heating and through-the-wall cooling; and (2) central oil-fired boilers for heating, domestic hot water and absorption cooling. Also, comparison was made with a total energy system that would provide heating to the Phase II buildings and electricity to both Phase I (now all electric) and the proposed Phase II.

North Town Phase II consists of 1000 residential units, limited commercial space and a small school. North Town Phase I has about 2100 housing units with baseboard resistance heating and through-the-wall cooling. Hence, this system was designated the reference system for the study. The concept includes piping Ravenswood Station condenser discharge water across the East River to central heat pumps on Roosevelt Island. The heat pump provides high-temperature water or steam to the buildings for space and water heating and chilled water for cooling in the summer. For the latter operation, East River water is withdrawn upstream of the island and the CE power plant, for cooling water input to the heat pumps. This study included preparation of a physical layout of the community thermal system with emphasis on the routing of an intake line from CE's condenser effluent and the cooling water from the East River to the heat pump site. The analysis included energy, economic and environmental comparisons between systems.

The feasibility study showed the following ranking by annual cost, with the lowest cost system first:

1. Heat Pump System I, using East River water without ice storage for air conditioning;
2. Heat Pump System II, using East River water with ice storage for air conditioning;
3. Heat Pump System III, using Ravenswood discharge water for winter heat but without ice storage for air conditioning in summer;
4. Heat Pump System IV, using Ravenswood discharge water in the winter with ice storage for air conditioning;
5. resistance electric heat;
6. conventional compression system;
7. total energy system; and
8. conventional absorption system.

Easton Utilities Commission (EU)

Easton Utilities provides electric power to a residential area surrounding Easton, Maryland, which is a rather sparsely populated area compared with the metropolitan areas of the U.S. Northeast. The Commission is not certain about promoting the use of heat pumps, even though a fairly large proportion of their customers already use electric resistance heating. However, it has been experimenting with water-source heat pumps. City water mains are serving as the external heat reservoirs for two water-to-air heat pump systems. The first installation in an 1800 ft^2 residence has operated through two heating seasons at a COP of 2.2, with water at 55°F. The second installation, on a "dead end" water main, was a failure. During cold weather, the water cooled down well below 48°, which would lead to compression failure from excessive loads. The flowrate of the water circulated through the successful heat pump is usually 6 gal/min (1590 liter/hr), with a temperature differential of 6°F (3.33°C) for a heat transfer of 18,000 Btu/hr (1.5 tons - 4500 kcal/hr). Electrical input is 4.5 kW for the heat pump and 0.75 kW for the water-circulating pump, leading to a COP of 2.2. For two heating seasons total power used by the heat pump averaged 6300 kWh. In addition, a backup resistance heater used about 1115 kWh per season because of compressor malfunctions. The Commission is trying to use a solar heat assistance to preheat city water when its temperature falls below 48°F. Four solar panels have been installed at the "dead end" heat pump.

Edison Electric Institute (EEI)

This is the largest U.S. association of investor-owned electric power operating companies and electric utility holding companies. It functions primarily in political and public relations areas working in conjunction with the Electric Power Research Institute for any special research projects. There are no Institute-wide programs to promote heat pump use or even any coordination of such programs being run by the member utilities.

Gulf States Utilities Company (GSU)

The Gulf States Utilities Company, operating in Texas and Louisiana, has been promoting the use of heat pumps for 25 years. Originally it offered a one-time subsidy in the form of a "wiring allowance" of $200 toward heat pump installation. At present, although GSU offers no service policies, the company will inspect the installation to advise the customer of any corrective measures that may be needed. There have been some difficulties due to careless installation, but in general, throughout the GSU system, only minor failures have been experienced. Under the prior policy of wiring allowances, the installation had to pass inspection to merit the payment, which resulted in fewer problems. The company feels that most problems are now due to dealer installation, rather than to inherent difficulties of the heat pump equipment as manufactured.

To demonstrate the efficiency of heat pumps and encourage more of their customers to have them installed, GSU has financed four projects to demonstrate the combined advantages of insulation, other energy-conservative structural features, and the use of heat pumps. Two homes in Beaumont, Texas, have been occupied less than a year and two homes in Baton Rouge, Louisiana have been occupied more than a year. The research work has been carried out by Texas A&M and Louisiana State University on primary contractors and preliminary results indicate energy savings of 30-35% for the whole system.

In a related project, GSU has sponsored Lamar University to study the advantages of thermal storage in mitigated demand peaks. For this purpose, a home has been built in Lumberton, Texas, according to GSU "Energy-Sensible Home" standards. It has a 1.5-ton commercial chiller with 292 gallons (1288 liters), or 2430 pounds (1102 kilograms), of ice storage in an outside, well-insulated, slab-mounted tank with submersed coils. Ice buildup is around the coils. Heat transfer is by water-to-air heat exchangers in the house. Circulating water is sprayed back into the tank. Heating is by a split, air-to-air heat pump with supplementary heaters.

The utility's interest in promoting heat pump usage is confined to its electric space heating customers. It cannot yet recommend it to its gas using customers.

National Rural Electric Cooperative Association

The National Rural Electric Cooperative Association has for its membership relatively small utilities that serve small, relatively isolated areas. Residential customers in these areas tend not to have gas available for space heating and must rely either on oil or electricity. Until now, the organization has had no strong program to promote heat pump usage. It usually suggests avoiding electric heating except for zone-controlled systems. Some research has been done on using a heat pump for hot water and also to assist in summer cooling. It anticipates bringing it on the market for about $400, in comparison with a solar-heated system of the same efficiency costing $2000.

Niagara Mohawk Power Company

The Niagara Mohawk Power Company serves a large, somewhat sparsely settled region in upstate New York. It has a fairly active program investigating the possibilities of advancing the heat pump state-of-the-art so that it can be profitably applied to its particular area, which usually experiences a rigorous winter. The program includes two different systems: one uses a water source with assist heat from a solar collector; the other is based on air. Both systems use electric resistance as a supplement. It also has sponsored a simulation study by Arthur D. Little on hypothetical systems representing the use of commercial systems for residence heating. It is primarily concerned with performance and is not related to cost targets.

The utility wants to have a valid technical base for design and evaluation of heat pumps suitable for northern climates. Specific hardware components are to be developed and tested for a heat pump suitable for the heating requirements in its service region. Eventually, the company should be able to specify the appropriate design for commercially available heat pumps that conserve energy in northern climate residential heating through good low-temperature performance and operating capacity. The dynamic computer simulation models have been used to identify factors affecting reliability and to study heat pump power demand profiles. A survey of candidate design and development possibilities for northern climates has led to four basic approaches being selected for detailed design and experimental verification. Field data analysis, which included reliability results, has been utilized in the selection of promising advanced heat pump candidates. The impact of cyclic operation and sizing in northern climates has been analyzed. Prototype advanced design units have been installed for testing.

The utility is also studying the use of a solar-assisted heat pump in a commercial building, aimed at determining the optimum thermal storage and operating mode for a combined solar-assisted space conditioning and water heating system within the constraint of minimizing power system peaking requirements.

Northern States Power Company (NSP)

This utility serves the metropolitan area of Minneapolis and St. Paul, Minnesota, probably the most difficult winter climate for heat pumps to combat. Nevertheless, the utility is promoting the use of heat pumps, primarily by distributing information on heat pumps and carrying out some experimental projects.

One current project is the Roseville House, which will require such little heat that a heat pump has a fair chance of meeting the winter heat requirement.

NSP has designed and built Roseville as a three-bedroom, octagonal, energy-saver home using dozens of conservation features. Not only is the construction of the home quite innovative, but heating and cooling is supplied by a modified Annual Cycle Energy System (ACES). The home will be occupied by a family to test its livability and will also be used as a demonstration for architects, engineers, manufacturers, government agencies and other utilities.

The two-story, octagonal design of the Roseville House affords maximum living space, yet reduces aboveground and wall and ceiling exposure for heat loss or gain. Placement of the house on the lot and the surrounding landscaping also help cut energy consumption. Berms are used around the house, and both deciduous and evergreen trees are strategically placed to protect the house from the elements. Insulation levels are even higher than those typically used in energy-saving homes: R-50 ceiling, R-26 first-floor walls, R-32 second-floor walls, and R-22 basement walls. The high levels of wall insulation are achieved through a combination of drywall, polyethylene vapor barrier, mineral wool, beadboard, plywood and cedar siding. Interior wiring does not pierce the vapor barrier in either walls or ceilings. Window area is limited to 10% of wall space, with a minimum for north-side windows. All windows are triple-glazed, with either interior or exterior thermal shutters, which increases the R-value to about R-8 or R-12. Insulated thermal break metal doors (R-12) and thresholds are installed. Both front and rear entrances have multiple doors that serve as airlocks when the inner and outer doors are opened and closed separately. The fireplace is located entirely within insulated walls, has an outside air intake, chimney cap damper and glass doors. A wood-burning stove in the basement may also be used for heating. Water use areas are centralized and all pipes are insulated. Flow restrictors are installed on all faucets and showers. Heat registers are replaced by heating air diffusers to reduce drafts. An Exhaust-Air Heat Recovery System uses vents and small motors to exchange household air with outside air, minimizing heat loss or gain.

A modified ACES system is used to heat and cool the house and includes a heat pump and 16,000-gallon low-temperature water reservoir located beneath the garage. Solar collectors on the garage roof supplement the ACES system in the winter by heating the low-temperature water reservoir. A 1000-gallon, high-temperature water storage tank is located in the basement. In the winter, the heat pump transfers heat from the 16,000-gallon reservoir to the 1000-gallon tank during the daily off-peak period while, in the spring, summer, and fall, the 1000-gallon tank is heated by the solar collectors. The water from the 1000-gallon tank is used to heat the house and preheat the domestic water. A 16-kW resistance heater serves as a backup and is used to determine the actual heating load of the house. There are five panels of monitors hidden in the walls, each containing 17 heat and humidity sensors. All applicances are also metered and monitored separately.

NSP is also experimenting with load management in the house. The ACES heat pump, backup resistance heater and electric water heater all operate off-peak. A small red light in the kitchen indicates peak and off-peak times. An electric meter separately records electricity used both on- and off-peak.

Pacific Power and Light Company (PPL)

This utility serves the outlying area near Portland, Oregon, and stretches to Wyoming which, in common with the rest of the U.S. Pacific Northwest coast, experiences annual climates that, in the entire U.S., probably most closely approximate those of Western Europe. Chapter 1 describes the narrow range between winter nighttime lows and summer daytime highs. While they have no active programs of subsidy or service contract, the utility does recommend heat pumps for new construction. A large part of the service area has some cooling requirement in the summer months. The biggest problem, because of the relatively mild temperature fluctuations, is to find a heat pump that is small enough. Sometimes the base load of electricity for cooking, illumination and refrigeration is enough for wintertime heat. PPL's experiments with heat pumps associated with the "TERA ONE" demonstration home began in June, 1977. Two water-to-water heat pump systems were tried, which were based on water storage in four tanks that had been heated by a solar air collector. The system marked well for heating, but not for cooling. The units had been custom made, incorporating an industrial compressor. There was a frequent requirement to recharge with the fluorocarbon refrigerant, due to excessive pressure differentials in operation. These heat pump systems have been replaced by air-to-air models, which have been working well.

Pennsylvania Power and Light Company

This utility provides electricity in a large stretch of central and eastern Pennsylvania, excluding Philadelphia. It does not have a program for promoting heat pump use other than fairly active experimentation in application of heat pumps in its service area. A number of houses were built and then modified to include a combination of solar cells and heat pumps operating on an air-to-air basis. It was felt that water-based heat pumps require very good control to avoid freezing the refrigerant under severe conditions. If part of the refrigerant is lost, the reduced charge allows the pump to cool, but it fails in heating operation. On basic operational data the utility is not pleased with the heat pump performance. The monthly and seasonal average COP fell below 1.5, instead of the range 1.8–2.0, as anticipated. Maintenance is still a problem. Only one of five systems in operation has required no maintenance. These were compressor failures due to refrigerant leakage, which led to excessive load and overheating.

To avoid this problem, all joints involved in the refrigerant circuit must be hard soldered (eutectic silver-copper alloys, m.p. 550-750°C). The utility found that the joints connected in the field installation frequently were not. Thermal storage was tried but most arrangements proved too expensive. It was suggested that water storage in a commercially available 100-gallon unit be heated at night using 6 kW to relieve the pump of working against the nighttime extreme. The objective of a current utility project is to reduce on-peak resistance heating as supplemental heat energy for the heat pump. Stored water heated off-peak is circulated through a water coil or heat exchanger. The utility, in its system, uses two hot water coils each for two roof-top heat pumps.

Portland General Electric Company (PGE)

The utility complements Pacific Power and Light by providing electric power to the metropolitan area of Portland. This area has abundant hydroelectric resources so electrical energy has always been cheaper than natural gas. The area's limits of hydroelectric development have been reached, however, and now the cost of electricity delivered to residential customers at 2.5 ¢/kWh (343 kcal/¢) is several times that of gas at 30¢/therm (100,000 Btu or 25,200 kcal). In new homes, 70% are still supplied with electricity for heating, which is marginally economic at present, but will not be as this area's electric rates rise.

In considering this problem, the utility has been experimenting since 1 February 1975, with solar-assisted heat pumps in selected customer residences. The aim is to gain some insight into operational characteristics of solar heating in this area, particularly in conjunction with heat pumps, as to its possible effects on the PGE system, and how solar power can be managed from a utility standpoint. The program has included installation of three different systems. The homes were designed to utilize "off-peak" electrical energy to supplement the solar contribution. To provide some guidance into the quantitative dynamic of the space heating problem, the system's operation has been simulated using the University of Wisconsin TRNSYS program. The homes have been instrumented with a 32-channel digital data logger to record performance in detail. PGE, in consultation with the University of Oregon Energy Solar Center, has produced educational brochures describing the program and promoting adaptation of solar heating and heat pumps.

The heat pump itself is water-to-air and seems to operate well down to 50°F water reservoir temperature, below which an electric resistance supplement meets the heating demand. Outside of a compressor being out three months for repair, the heat pump has been essentially maintenance free. The heat pump and solar collector do complement each other in operation; however, the mild climate places so little demand for space heating that the combined investment is not yet economical. Recently, the utility has been experimenting with wood burners for supplemental heat to the water reservoir.

Potomac Electric Power Company (PEPCO)

This utility provides electric power to Washington, D.C., and the nearby metropolitan area in Maryland. The same area is served with gas by

Washington Gas and Light Company. Most homes have been heated by gas because it was cheap. Since the weather extremes are mild most years, homes in the area tend to be poorly insulated. With recent uncertainty over gas availability there was a movement among utility customers having both services to install heat pumps as an alternative, with the option of using gas during low-temperature extremes. This created some antagonism between the two utilities. PEPCO has a low-key program to promote heat pump installation among its customers who are already on electric heating.

The utility has also sponsored three years of experiments carried out by Westinghouse on the residential use of a solar-augmented heat pump with thermal storage. The project had the following objectives:

1. At no degradation in lifestyle, demonstrate a space comfort-conditioning system using scheduled off-peak operation of the system for both heating and cooling. The system must incorporate only state-of-the-art components in solar collection, heat pump and thermal storage units.
2. With comprehensive instrumentation, obtain necessary data to evaluate performance, efficiency and load profile in a typical field installation.
3. Using the information obtained, compare the impact on the utility load factor of this type of system to that of conventional systems as a function of market penetration.

Public Service of New Mexico (PSNM)

This utility provides electric power over a wide area where the climate is warm enough that the principal requirement is cooling during the summer. As a result, the utility load is summer-peaked. Since the state is also an area of high insolation, a major research and development interest of the utility is in the solar "power tower" concept to provide supplemental solar-generated electricity during the summer. Nevertheless, PSNM's interest in conservation is strong enough to have collaborated with EPRI and a group of utilities in the testing of heat pumps by the Westinghouse Corporation (Blake and Gernert, 1978) primarily oriented toward usage in the northern states.

The utility monitored 10 homes with heat pumps within the Albuquerque area. Information has been used to compare operational characteristics of each heat pump unit as a function of weather conditions. Anticipating that many new homes will utilize heat pumps, PSNM wants to establish guidelines for expected operating and maintenance (O&M) costs, installation problems, etc. At each residence the following environment and operational parameters were recorded:

- indoor dry bulb temperature;
- energy use by the indoor fan and resistance heater; and
- compressor operation in both heating and cooling modes.

At the weather station, outdoor dry bulb temperature and direct and diffuse components of solar radiation were recorded. In a correlative experiment, the University of New Mexico subjected a heat pump to tests in a special environmental chamber at UNM.

SPF analysis (performance over an entire year) was done for 5 of the 10 homes. They were found to have an average SPF of 1.41, which represented a savings of 29% over an equivalent electric resistance furnace. Another 4 homes were analyzed for demand. Individual load factors were found to be lower for the heat pump than for the resistance mode, as expected. However, the system coincident load factors were typically higher for the heat pump than for the resistance mode. This would indicate that commonly experienced weather extremes imposed a synchronized load on the heat pumps that would have been randomized out (due to internal factor variation) for resistance heating.

Public Service Electric and Gas Co. (PSEG)

This utility serves most of New Jersey in the northern half of the state. Although it sells both gas and electricity, the utility promotes the use of air-to-air heat pumps among its customers who already use electricity for heating. They have been issuing informational brochures but no service or guarantees.

The utility is also cosponsoring a program in conjunction with the Association of Edison Electric Illuminating Companies to look into thermal energy storage. The work is part of an expanded heat pump investigation being done by the Princeton University Center for Environmental Studies, seeking to improve efficiency and seasonal operation of major appliances. The Center is investigating thermal energy storage for use in load leveling and to improve load factors.

Southern California Edison (SCE)

This utility serves a wide area centered on San Francisco. The climate is mild, but there is some need for heating in winter. SCE is promoting solar installation quite heavily because of the demand motivated by the large combined federal and California subsidy for solar equipment. It does not necessarily promote heat pumps, but does get interested in those systems incorporating a solar assist. The current project sponsors, the Jet Propulsion Laboratory and Blurock and Partners, are studying the utility of a solar-augmented heat pump for the KOCE-TV office building. The four-year project was begun 1 March 1976, funded for $230,000. The aim was to demonstrate and obtain engineering data for a solar-augmented hydronic heat pump system in a commercial building. Separate objectives were to:

1. develop and validate design procedures and assumption for solar augmentation of heat pumps in commercial buildings;
2. demonstrate the energy conservation potential of solar-augmented heat pump systems;
3. collect data for future optimization of the system for application in customer buildings; and
4. develop a design and sizing for solar hydronic heat pump systems in residential buildings.

A precedent study, also done by the Jet Propulsion Laboratory (French et al., 1976), studied the design and economics of solar-augmented heat pumps with thermal storage for Southern California homes. Based on their analysis and that of Davis and Wen (1975), French et al. concluded that the most cost-effective heating, ventilating and air conditioning system and the best application of solar energy for an all-electric home in Southern California is a conventional heat pump plus solar heating of domestic water. In an energy-conserving house, the payback period for the heat pump, when compared to resistance heating, will be about 6 years. The payback period of a solar water heating system, if produced on a commercial scale, has been estimated (Davis and Wen, 1975) to be about 10 years. The concept of system studied is shown in Figure 5-1. A summary of the study's results is shown in Table 5-4.

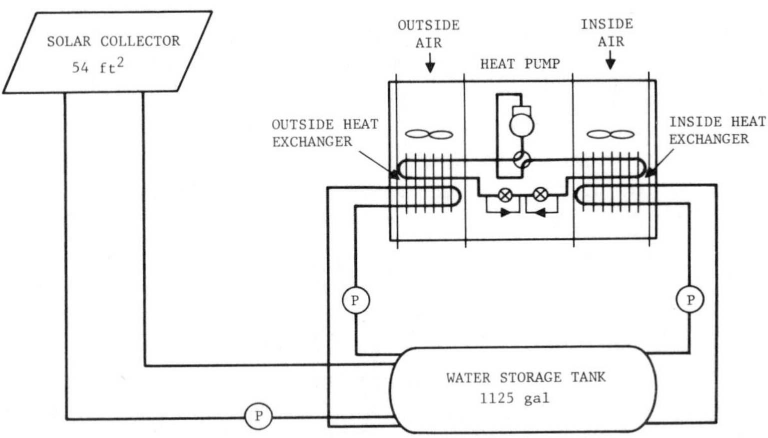

Figure 5-1. Solar-augmented heat pump system with thermal storage (Southern California Edison).

Table 5-4. Solar-Augmented Heat Pump with Thermal Storage Model Results[a]

System	Comments
Off-Peak-Power Cooling	
Cold water storage	12% increase in electrical energy consumption. 7-year payback if day-night electrical differential exceeds 0.035 $/kWh.
Ice storage	32% increase in electrical energy consumption. 7-year payback if day-night electrical differential exceeds 0.003 $/kWh.[b]
Heating	
Heat pump (current)	Baseline, SCOP = 2.4.
Solar system in series with heat pump	18-year payback.
Solar system in parallel with heat pump	13-year payback.
High-performance heat pump (future)	SCOP[c] = 3.4, 13-year payback if incremental cost is $220/ton; 19-year payback if incremental cost is $330/ton.

[a]All cost estimates are based on 1974 dollars.
[b]Cost of electricity is assumed to be 0.035 $/kWh.
[c]SCOP is the Seasonal Average Coefficient of Performance.

Texas Power and Light Co. (TPL)

This utility supplies electricity to large metropolitan areas in Texas, notably the Dallas-Fort Worth area. For many years, similar to the Alabama Power Co. arrangement, TPL offered heat pump maintenance policies in which service was guaranteed. This program was dropped a few years ago when the utility felt that heat pump use in the area had become accepted as a routine option. TPL's service experience was fairly good, with compressor burnouts being relatively rare.

At present, the utility is concentrating its effort on a demonstration program, "The Solar Mate Home–A Heat Pump-Solar Collector Interface," which incorporates a new developmental two-speed compressor heat pump model being introduced by Lennox. A particularly novel feature is the use of solar-heated water to directly heat the refrigerant in the compressor.

Along with TPL, Design Builders, Inc. and Lennox Industries, Inc. are participating as co-sponsors. Their demonstration is to be a practical experiment, which could be duplicated with minor modification to represent any climate and building structure in the U.S. A standard structure and mass-produced, reasonably priced solar collectors were used. They expected that the system used in this experiment would pay for itself within the five-year warranty period wherever electricity costs exceed 6¢/kWh. The demonstrator house was well sealed and insulated before installing the system which is rated high cooling EER (8-10) and heating COP (3-5). The COP greater than 3 was expected to be achieved by the solar assist in heating the refrigerant and domestic hot water.

In overall performance of the system schematically shown in Figure 5-2, the only disappointment was the solar collector, which provided only 8%

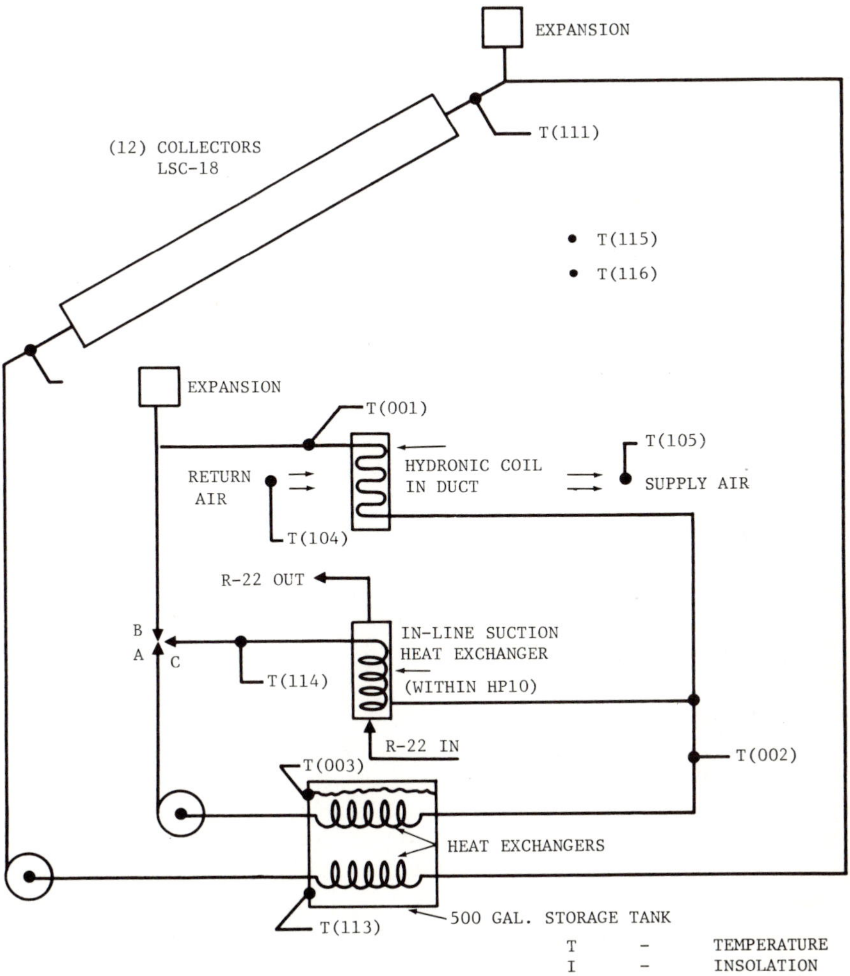

Figure 5-2. Flow diagram for solar-assisted two-speed heat pump in Dallas, Texas (Texas Power & Light Company).

of the total heat supplied to the home. Overall system performance was a COP of 1.75. The two-speed heat pump itself had an average compressor COP of 2.54. Heat pump performance was influenced strongly by two interesting features. The unit needed a defrost cycle only when the solar assist was removed, which may have been the major benefit of the solar assist. Also, the high-speed mode provided 83.6% of heat pump heating, while 16.4% was at low speed. Hence, primarily low COPs associated with high speed, as shown in Table 5-5, were reflected in the heat pump results. It is possible that this distribution between high- and low-speed operation could be skewed toward more low-speed operation for higher COP by modifying the programmed control system. Limits on cross-compressor pressure differential would have to be accommodated in the control.

Table 5-5. ARI Ratings (designed) for Lennox HP10-511 with CBP10-65 Two-Speed Heat Pump (4 ton/2 ton = 48,000/24,000 Btu/hr)

COP		EER, Btu/W-hr
Cooling[a]		
High Speed	2.5	8.4
Low Speed	2.8	9.6
High-Temperature Heating[b]		
High Speed	2.9	9.9
Low Speed	3.3	11.3
Low-Temperature Heating[c]		
High Speed	2.1	7.2
Low Speed	2.4	8.2

[a] 95°F db outdoor; 80°F db, 67°F wb indoor.
[b] 47°F db, 43°F wb outdoor; 70°F db indoor.
[c] 17°F db, 15°F wb outdoor; 70°F db indoor.

Tennessee Valley Authority (TVA)

Established during the 1930s to distribute electricity over a wide area embracing Tennessee and parts of Kentucky and North Carolina, TVA is a federal entity. It does not address its annual report to its customer, as a municipally owned utility does, but rather to the President of the United States, the Senate and the House of Representatives. Its employees number 40,000 and its utility customers 7,000,000. Historically, its rates have been cheap, based on hydroelectric power; also, the region's climate is considerably milder than the northern tier of the U.S. Hence, there is a great deal of electric heating. Lately, however, it has had to use more fossil fuel and rates are climbing. The Authority is now starting a program to promote the use of heat pumps with a certification program similar to that of the Alabama Power Co., including some financial subsidy. Financial assistance will consist of low-interest loans (expected 8½% at present), in any reasonable (expected \$2000-\$4000 range) amount required for purchase and installation. Loan repayments will be assessed on the monthly electric bill amortized up to 10 years. TVA standards determine whether the heat pump installation is eligible for the loan.

The Authority is also sponsoring a small solar-assisted heat pump demonstration. The stated objectives are to determine the economic feasibility of using solar energy to supplement electric heat pumps for residential space heating and domestic water heating on the TVA system. TVA is interested in evaluating systems for demonstration that may be able to reduce system peak demands and conserve energy. It estimates on its heat pump promotion program, for example, a cost to TVA of \$6 million, with \$12 million savings to the power system.

Electric Power Research Institute (EPRI)

EPRI serves to coordinate and sponsor, in part, most of the electric power-related research for utilities in the U.S. The subject of heat pumps has received increased attention recently, although most of the projects seem to involve some solar energy assistance. This can backfire, however, since, even at the demonstration stage, the solar components often incur delays in construction and installation while the heat pumps experience no problem. One example is a rather ambitious demonstration reported by Stewart et al. (1977) using the Alumni House-Conference Center (AHCC) on the campus of the State University of New York at Albany, which is in a region that characteristically experiences a very severe winter. The heating system combines 2300 ft^2 of copper solar panels with a water-to-water heat pump, an electric boiler and two 8000-gallon storage tanks. The building is instrumented to provide data for the time variant electric power demands, the impact of meteorological variables, peak storage capabilities, the efficiency of the solar panels, coefficient of performance of the heat pumps and the heating system performance factor. Computer simulations of the heat budget of the AHCC have been prepared and verified. A series of economic analyses provide Return on Investment, comparing the air-source heat pump to the solar-assisted water-to-water heat pump and using a base case of electric heating peak shaving.

Unverified computer analysis indicates that the air-source heat pump has a higher COP than the solar-assisted water-to-water heat pump. However, during peak winter operating conditions, the solar system provides 33% of the heating load of the AHCC. The overall purpose of the study was to monitor the performance of the integrated solar energy and heat pump system of the AHCC of the SUNY at Albany. The individual tasks originally assigned were to:

1. internally monitor the performance of individual system components;
2. analyze results of (1) to suggest modifications for improving the efficiency of the system;
3. implement the modifications suggested by (2) if possible, to be done only in consultation with the Niagara Mohawk Power Corporation;
4. reevaluate the system after modification;
5. provide (from present stock) meteorological instrumentation to monitor weather conditions on the Alumni House site;
6. analyze all necessary meteorological variables affecting the availability of the solar energy system to function in this climate;
7. forecast weather conditions 24-48 hours in advance to optimize the use of the solar energy storage system, thus reducing peak loading;
8. provide an economic analysis of the system based on initial and operating costs over the two-year project period;
9. compare the operation cost of this structure to other buildings of similar size using conventional heating/cooling systems;
10. pinpoint procedures for decreasing the overall cost of operating the structure by reducing total energy demand, or by smoothing the pattern of usage and thus setting up a procedure for reducing the peak load demand on the utility;
11. analyze the implications suggested by the results of the economic and physical studies of 1, 2 and 3 above; and
12. suggest the direction for future research efforts in this area of solar energy.

Not all these tasks were completed. The overall program was terminated short of the final contract date due to a series of difficulties, especially delays within the construction program. The solar system was tested during the winter of 1975-76, but did not become operational until November 1976. Since that time, the system has performed without incident.

The AHCC provided the ability to test a solar heating system, a solar-assisted water-to-water heat pump system, an electric boiler, an air-to-water heat pump system and peak shaving, all within the same structure. Data collection capability included 60 channels of information on flat plate temperatures, flowrates, electrical demand by pumps, motors, fans, etc. These data were used to verify the simulation of the various systems. Based on these simulations, the AHCC would require 79,930 kWh/yr for heating, if only the electric boiler were operated. The same heating load would be met by the solar-assisted heat pump, consuming 50,780 kWh. The total kWh predicted for the air-to-water heat pump would be 45,160 kWh. While the first two demands have been verified using operational data, the predicted kWh usage of the air-to-water heat pump is based on manufacturer's specifications. However, the impact of this result cannot be ignored and must be verified.

During a 29-day winter test period, the solar collection system provided 33% of the heating demand (the period had 13% more degree days than normal but received 5% more solar radiation than normal).

The energy and economic analysis indicates that the air-source heat pump is attractive both on return on investment (ROI) and on minimizing electric demand. The base case all-electric system cost \$45,000, the solar-assisted heat pump \$97,700 (\$52,700 for the solar panels, etc., plus \$45,000 for the electric backup), and \$58,000 for the air-source heat pump. The additional capability of peak shaving required an additional \$11,400. Analysis indicated that this solar-assisted heat pump will begin to show a return on investment when the cost of electricity approaches 14¢/kWh for a corporation, or 10¢/kWh for a municipality. Further sensitivity analysis is also to be performed. Heating costs per square foot in the Albany region are also included for comparison to the AHCC. For offices in this region, heating costs range from 30-70¢/ft^2. At 4.5¢/kWh, the solar heat pump system would heat the AHCC at 30-35¢/ft^2/yr.

Heating Demand Analysis

The simulated heat budget for the AHCC is presented in Table 5-6. Table 5-6a provides the analysis based on the electric boiler only. Thus, if only the electric boiler were operational, the AHCC would require 79,930 kWh/yr for heating. Table 5-6b provides the same analysis, but based on operation of the solar-assisted heat pump. While the electric boiler still receives substantial usage, annual kWh consumed for heating decreases to 50,780. Table 5-6c provides the analysis of the air-to-water heat pump, which was added onto the project. Total kWh predicted for this unit are 45,160. Data for the first two units (electric boiler and solar-assisted heat pump) have been verified. The air-to-water heat pump data were not verified before the project was terminated by EPRI. This point is critical as this initial analysis indicates that the air-to-water heat pump is more efficient in meeting the heating load than either the electric boiler or the solar-assisted heat pump.

The economic results for these simulations are shown in Table 5-7, which indicates that the air-source heat pump is attractive both in terms

Table 5-6. Simulated Heat Budgets for Alumni House Conference Center, Projected for a Heating Season (Stewart et al., 1977)

a) Electric Boiler

Month	Average Ambient Air Temperature (F°)	Heat Supplied to Load[a] Electric Boiler	Heating System[b] (kWh)
Sep	61	0.8	2,330
Oct	56	4.7	3,550
Nov	42	22.7	8,750
Dec	32	40.2	13,950
Jan	28	48.1	16,260
Feb	30	39.5	13,530
Mar	35	34.5	12,280
Apr	50	12.7	5,820
May	58	4.4	3,460
Totals		207.6	79,930

b) Solar-Assisted Heat Pump

Month	Solar Energy Collected (10^6 Btu)	Heat Supplied to Load			Heating System[b] (kWh)
		Direct Solar (10^6 Btu)	Heat Pump (10^6 Btu)	Electric Boiler (10^6 Btu)	
Sep	4.5	0.8	0.0	0.0	2,220
Oct	6.6	4.7	0.0	0.0	2,250
Nov	12.3	2.2	18.7	1.7	4,430
Dec	13.1	0.0	19.0	21.2	10,370
Jan	17.8	0.0	20.7	27.4	12,360
Feb	21.7	0.0	29.4	10.1	7,880
Mar	24.7	0.0	34.0	0.6	5,600
Apr	17.3	2.0	10.0	0.7	3,390
May	8.8	4.4	0.0	0.0	2,280
Totals	126.8	14.1	131.8	61.7	50,780

c) Air-to-Water Heat Pump System

Month	Average Ambient Air Temperature (°F)	Heat Supplied to Load		Heating System[a] (kWh)
		Heat Pump (10^6 Btu)	Electric Boiler (10^6 Btu)	
Sep	61	0.8	0.0	2,160
Oct	56	4.7	0.0	2,540
Nov	42	22.7	0.0	4,050
Dec	32	37.4	2.8	7,400
Jan	28	40.0	8.1	10,900
Feb	30	36.5	3.0	7,220
Mar	35	34.5	0.0	5,260
Apr	50	12.7	0.0	3,130
May	58	4.4	0.0	2,500
Totals		193.7	13.9	45,160

[a]Seasonal heating load = 207.6 x 10^6 Btu.

[b]Includes pump and fan kWh.

Table 5-7. Energy and Economic Results of Alumni House Simulation (Stewart et al., 1977)

Energy Distribution and Financial Incentive	Energy-Conserving System: Air-Source Heat Pump (%)	Energy-Conserving System: Solar-Assisted Water-to-Water Heat Pump (%)
Heat provided by system, %	93	78
Heat provided by electric backup, %	7	22
Electric generation saved, %	63	55
Return on investment for a corporation	7	None
Return on investment for a municipality	22	None

of energy savings and financial incentive. There is a 63% reduction in the amount of electricity, which must be generated by fossil fuels. In addition, there is a financial return of 22% over a 30-year bond-financing period at 7% interest. That is, a municipality would more than recover the added cost of a heat pump through heating cost savings. A corporation funding the building's heating system with 40% of the company's equity and the remaining 60% borrowed at 10% interest over a 25-year financing term would realize a 7% return on investing in the heat pump, rather than in a conventional electric heating system. This 7% would be increased through a longer period of fuel cost savings if the life of the heat pump is assumed to be 30 years, as in the municipal financing case.

The solar-assisted heat pump, while exhibiting substantial energy savings, did not prove economically viable at the 4.5¢/kWh cost of electricity currently in effect in the Albany area, when joined with the Alumni House solar system costs of $\$23/ft^2$.

A cost breakdown for the air-to-water heat pump system itself is presented in Table 5-8.

Difficulties and Delays

As was mentioned earlier, the expected schedule for completion of the AHCC was not met, due mostly to the fact that a firm completion date on the building was not enacted. Although most of the delay was due to the general construction schedule, there were specific delays due to the solar system. The most serious problem was created by a series of separations in the copper piping, which connected the solar panels to the header. This was due to the expansion of the copper tubing during dry stagnation.

Table 5-8. Air-to-Water Heat Pump System Costs (Stewart et al., 1977)

Component	Cost ($)
Two 12-ton air-to-water heat pumps	12,000
Four 20-kW electric boilers	2,000
One 275-gal tank	300
Three air-handling units	3,500
Seven reheat coils	3,535
HVAC piping costs	7,000
Ducting, louvers, grills, diffusers, registers, etc.	10,000
Domestic hot water heater	500
Temperature controls	6,000
Electrical	3,000
Design engineering	2,810
Total costs	55,010
Total costs plus 15% for overhead and profit on all items except design engineering	57,820[a]

[a]The final total cost was rounded to $58,000 for use in the economic analysis.

The final solution to the problem was the installation of a flexible metal base (Flexonics Hose Type 300), which allowed for expansion of the header and/or panels without requiring a rigid connection between the two.

While this was the most serious problem of the solar system, other minor delays were necessitated by the need to check and recheck modification during construction. There were also some weather-related delays during the assembling of the solar panels on the roof (rather than using modular units), which required more time and, thus, more exposure to the weather. Eventually, EPRI terminated the program.

U.S. GOVERNMENT AGENCY PROGRAMS

U.S. Government-sponsored heat pump system research is carried on primarily by DOE and EPA. There is no such current work identifiable at the Departments of Commerce (DOC) and Housing and Urban Development (HUD), the only other two agencies that would have a special interest.

EPA's program is much smaller than DOE's and is administered through Research Triangle Park. The Agency's interest originated because the use of heat pumps promised to reduce demand for premium fuels. Since then, a special studies group has been studying heat pump utilization from the standpoint of energy savings. EPA is interested in the possibility of reducing requirements for growth in electric power generation capability. However, serious reservations about fuel and heat pump savings have been expressed. Currently, the only EPA investigative areas relevant to heat pumps are the following, for which no reports are available at present:

1. heat pump as a substitute for outmoded fossil systems; and
2. coal-fired furnaces vs heat pumps for residential heating.

In these studies, both environmental and economic issues are being considered. About 100 residences are being analyzed for environmental impact, but without considering the effect—a refrigerant leakage to atmosphere with possible reduction of upper ozone levels. This issue would have to be addressed before they could issue a final report.

DOE has a stronger program because it has a direct interest in energy savings as a fundamental feature. At present, its sponsored research is administered out of the Technology and Consumer Products Branch. Outside of the cosponsorship with EPRI of the utility-centered program already outlined in this chapter, DOE's research is done mainly by a fairly strong group at the Oak Ridge National Laboratory. An allied branch, federal programs (for federal buildings), has no sponsored research, but is attempting to establish standards for federal building heat pump use. Some information should be forthcoming soon to support the standards. Both branches are under the divisions of Buildings and Community Systems of DOE.

Generally, DOE takes two contrasting attitudes toward energy research. On the one hand, there is intense interest in what is possible now, and elaborate programs are organized to establish exactly what is the current state of technology. In these studies, any prospects beyond off-the-shelf equipment are considered irrelevant. On the other hand, they are also interested in major technological advances, in which concern over economic uncertainties or near-term developmental feasibility is deferred, at least temporarily. The utility programs are aimed in between to ascertain their next step.

An extremely interesting example of an advanced heat pump system experiment is a demonstration of what can be achieved through comprehensive use of annual thermal storage (Holman and Brantley, 1978). A 2000-ft^2 single-family residence was constructed to demonstrate the energy-conserving features of additional insulation, a ventilation cooling cycle and the Annual Cycle Energy System (ACES).

The latter denotes the concept of annual thermal storage similar to that of the "Roseville House" Project of Northern State Power. ACES is an integrated heating and cooling system that supplies space heating, hot water, and space cooling using a heat pump and a more elaborately designed low-temperature thermal storage. The report discusses the construction techniques employed, the problems encountered during construction, the ACES concept, the ACES mechanical package, the ACES performance and the experiences obtained during its initial operation. Continuous operation and data collection were begun in May 1977. Performance data from November 1977 through March 1978 have shown a seasonal heating coefficient of performance of 2.77 for the ACES, giving a 63% energy reduction compared to electric resistance space and water heating.

In the Oak Ridge ACES, the outside air unit of the heat pump is replaced with ice-freezing coils located in an ice-and-water thermal storage bin, and the external source is the heat of fusion of water as it is converted to ice.

During the heating season, as the compressor operates to supply the heating demands, water is converted to ice in the thermal storage bin. When space cooling must finally be supplied, the ACES need only melt stored ice to meet the cooling demands. Because the compressor does not operate to meet the cooling loads, energy savings are obtained over a conventional heat pump system. Additional energy savings are obtained by producing the domestic hot water with the compressor at a COP of approximately 3, as opposed to using electric resistance water heating at COP of 1. The ACES does not require supplemental resistance heaters, so that it can deliver all of the space heating at a nearly constant COP of 2.7, even during the coldest periods of winter.

Because the ACES employs a finite heat source, as opposed to the nearly infinite heat source used by a conventional heat pump system, some mechanism must be provided for balancing the ACES over a full cycle of seasons. A single outdoor radiant/convector panel is used either to accept or reject heat. If, during the heating season, too much of the water in the thermal storage bin is being converted into ice, the solar panel can be used to accept heat from the sun (and/or the ambient air if the temperature is in excess of 32°F) to melt some of the ice. If, on the other hand, the ice supply in the thermal storage bin has been exhausted before the end of the cooling season, the panel can be used at night to reject heat from the compressor. This nighttime operation of the compressor, which produces ice in the thermal storage bin for later use in space cooling, provides off-peak load leveling, even when the winter ice supply has been exhausted.

The ACES, shown in the schematic flowsheet of Figure 5-3, is composed of a compressor, four refrigerant-to-brine heat exchangers, a fan coil, an outdoor panel, a domestic hot water storage tank, a thermal storage bin and the appropriate piping and valving to interconnect the various components. The "brine" in the ACES is a solution of 25% methanol by volume in water with freezing point depressed to around -20°. By employing various valving arrangements, the ACES has nine distinct modes of operation: three

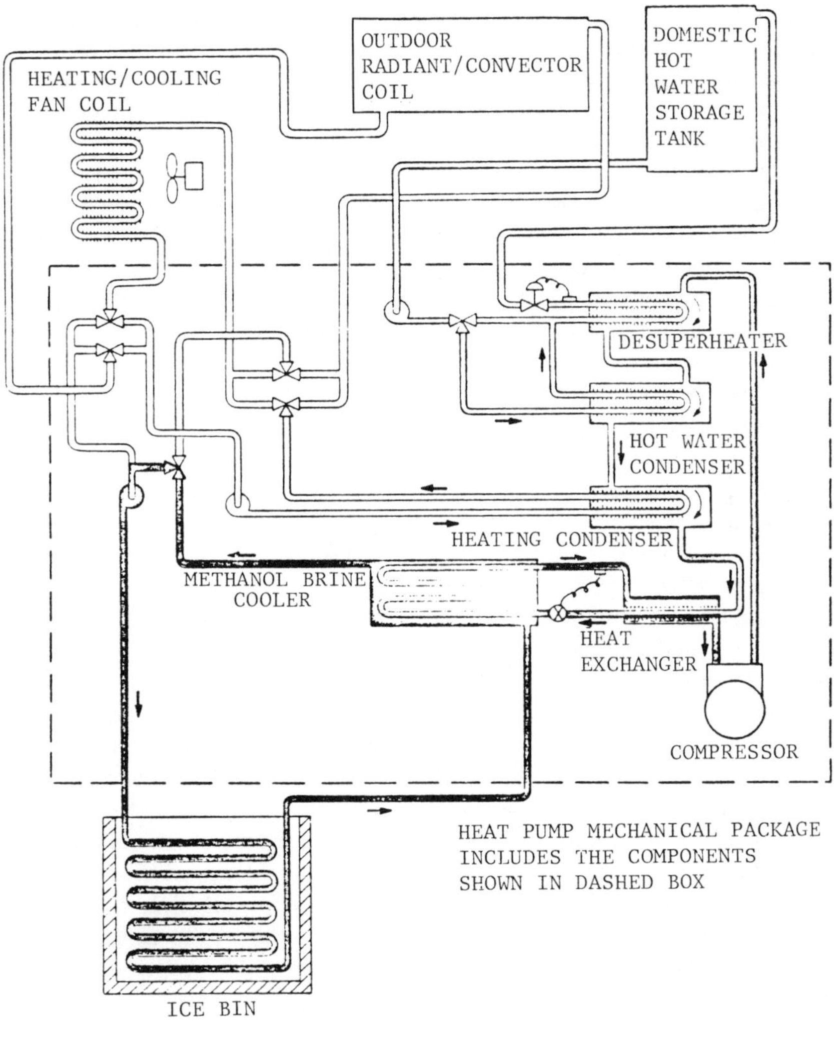

Figure 5-3. ACES space heating and cooling (Holman and Brantley, 1978).

perform the primary functions of space heating, space cooling and domestic hot-water production; two perform the balancing functions of ice melting and night heat rejection; the remaining four modes result from combinations of such modes occurring simultaneously.

Several other relevant DOE investigations, all carried out at Oak Ridge National Laboratory, are in the other category of systematic, rather exacting reviews of the state of existing technology and its capabilities. Unitary air-to-air heat pumps were surveyed by Christian (1977). The report covers a large capacity range, from 1.5-45 tons. Seasonal COP depended on climate, of course, and varied from 2.5 down to a startlingly low 1.1. The latter figure represents complete failure, since capital and maintenance for such a heat pump are for a negligible energy gain over resistance heating. The energy difference could even be negative, considering the loss of resistance heating control flexibility. In another study, an extraordinarily comprehensive and detailed evaluation of a heat pump (capacity of 3 tons = 9040 kcal/hr) was carried out by Domingorena (1978). In addition to the usual superficial measure of performance, the underlying physical dynamics of the heat exchangers, capillary tubes, fans and compressors were exhaustively instrumented and observed for both normal and defrost operation. Overall results were as discouraging as Christian's work, promising a seasonal performance factor between 1.5 and 2.0.

The difficult problem of rigorously testing air conditioners and heat pumps operating as coolers was addressed by Kelly and Parken (1978), working out of the National Bureau of Standards (NBS) but funded by DOE. They looked into both front-load and seasonal performance, outlining methods for testing and rating that account for those variations in performance due to part-load operation and changes in outdoor temperatures (heat rejection conditions).

The use of heat pumps for water heating was compared to other means by O'Neil et al. (1978). The analysis was regional and envisioned some scenarios in fuel price variation to the year 2000. The systems compared are: (1) electric heat pumps offered in 1981; (2) solar systems offered in 1977; and (3) solar systems offered in 1977, with a federal tax credit in effect from 1977 through 1984. A residential energy model developed by the Oak Ridge National Laboratory (ORNL) is used to calculate energy savings by type of fuel for each system in each of the 10 federal regions and for the nation as a whole for each year between 1977 and 2000. Changes in annual fuel bills and capital costs for water heaters are also computed at the same level of detail.

Model results suggest that heat pump water heaters are likely to offer much larger energy and economic benefits than will solar systems, even with tax credits. Heat pumps provide about the same savings in electricity for water heating (about half) at a much lower capital cost ($700-2000) than do solar systems. However, these results are based on highly uncertain estimates of future performance and cost characteristics for both heat pump and solar systems. The cumulative national energy saving by the year 2000 due to commercialization of heat pump water heaters in 1981 is estimated to be 1.5 QBtu. Solar energy benefits are about half this without tax credits and two-thirds as much with tax credits. The net economic benefit to households of heat pump water heaters (present worth of fuel bill reductions less the present worth of extra costs for more efficient systems) is estimated to be $640 million. Again, the solar benefits are much less.

This anticipation of commercial possibilities for heat pump water heaters superior to solar energy systems is apparently being fulfilled by the Temcor unit as described by Gorzelnik (1979). There are now a number of utilities listed in Table 5-9 that are formally committed to the use of five Temcor units each. The Temcor has been developed by Energy Utilization Systems, Inc. of Pittsburgh, Pennsylvania, with substantial DOE sponsorship, and full production of 82-gallon units was anticipated by September, 1979. These units are to be sold to utilities at $371 each. A heat pump add-on section to retrofit existing water heaters would sell for $200.

Table 5-9. Utilities Participating in Heat Pump Water Program

Arizona Public Service, Phoenix, Arizona	Somerset REC, Somerset, Pennsylvania
BPA, Portland, Oregon	South Carolina E&G, Columbia, South Carolina
Florida P&L, Miami, Florida	Southern California Edison, Rosemead, California
Gulf Power, Pensacola, Florida	TVA, Chattanooga, Tennessee
Hawaiian Electric, Honolulu, Hawaii	Valley Rural Electric Co-op, Huntington, Pennsylvania
Indianapolis P&L, Indianapolis, Indiana	
Kansas Electric Co-op, Topeka, Kansas	**In Negotiations**
Mississippi P&L, Jackson, Mississippi	Duke Power, Charlotte, North Carolina
New York State E&G, Binghamton, New York	Guadalupe Valley, EC, Gonzales, Texas
Portland GE, Portland, Oregon	Kansas G&E, Wichita, Kansas
Public Service Indiana, Plainfield, Indiana	

The Temcor water heater shown in Figure 5-4 consists primarily of a vapor compression engine mounted on top of the water with very simple controls. It is designed to extract heat from the ambient air surrounding the evaporator and convey it to a condenser coil immersed below. Ambient air temperature limits for a COP of 3.0 are 45–95°F. The unit cannot tolerate any frosting of the evaporator or refrigerant gas pressure in the condenser above 300 psi. A heated basement or unheated garage in very moderate climates are then suitable locations. It is likely that in winter the unit can use waste heat from the furnace. In summer, it will simply reduce overall air conditioning requirements.

One of the most advanced fundamental heat pump concepts that promises both operational simplicity and a 50% improvement in COP was

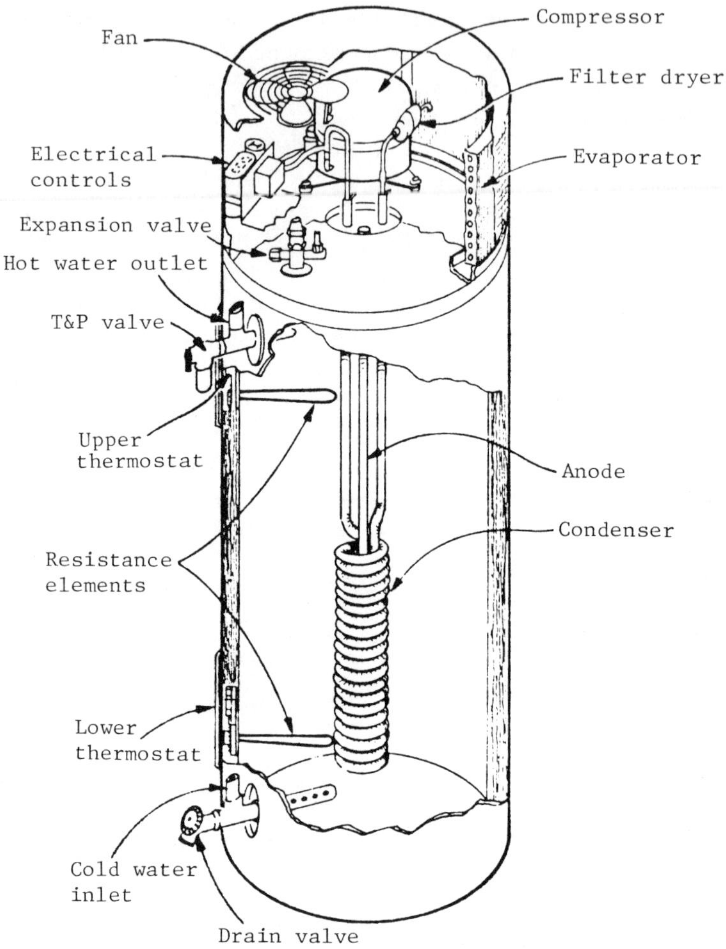

Figure 5-4. Temcor heat pump water heater (Gorzelnik, 1979).

developed in 1978 by Gerald Brown of NASA's Lewis Research Center in Cleveland. He showed that the magneto-thermal effects in gadolinium can be used to establish temperature differentials of 140°F between heat exchangers operating at ambient temperature levels. The system operates on the inherently reversible cycle of heating and cooling when a magnetic field is applied and then reduced for a highly permeable substance. Typically operational levels of at least 60% of those of an ideal Carnot engine are achieved compared to the 40% or less of the typical vapor-compression cycle. At present, the economics are not very favorable due to high cost of gadolinium and a high field magnetization system. This necessitates a high-capacity unit suitable for industrial or commercial use. However, gadolinium is relatively abundant and, with a market, may be produced cheaply. The expensive magnetization system may then be replaced by inexpensive use of a permanent magnet for a low-capacity system that is extremely simple and, therefore, may be both reliable and durable. One of Brown's conceptual designs is illustrated in Figure 5-5. The most recent descriptions of Brown's magnetic heat pump principles are in a

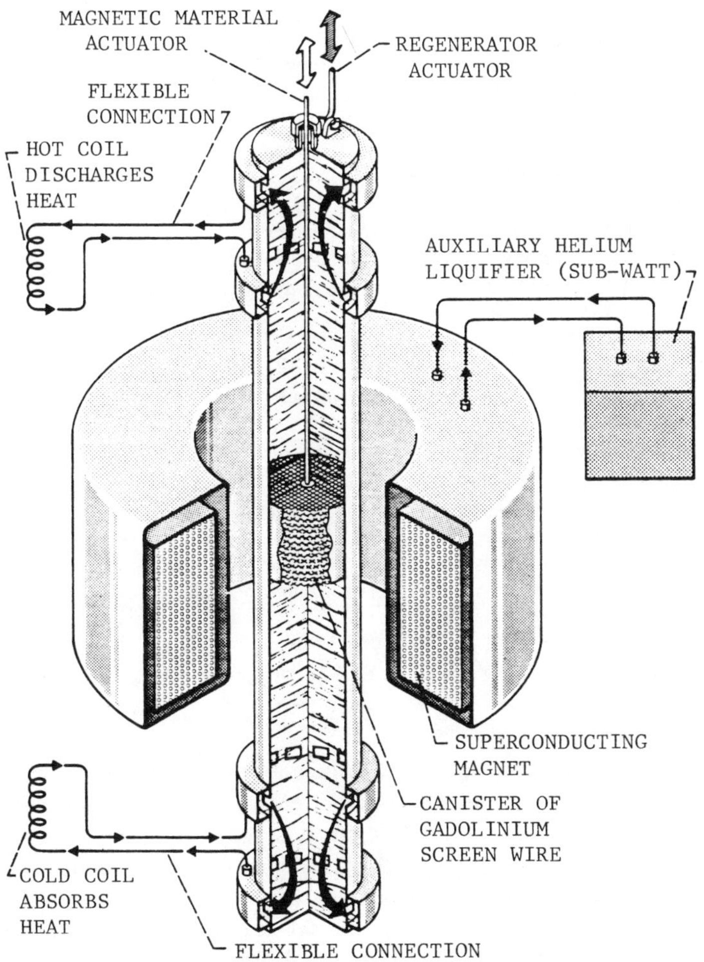

Figure 5-5. Magnetic heat pump (Brown, 1978).

patent (Brown, 1978) and two articles (Brown, 1976; Brown, 1977) on magnetic heat pumping without a supercooled magnet.

TAX BENEFITS AND OTHER SUBSIDIES

The only direct subsidy ever offered for heat pumps was the now terminated "wiring allowance" mentioned in describing the Gulf States Utilities program. At present, there are no federal or state direct subsidies. However, in the new U.S. Energy Tax Act of 1978 (Public Law 95-618, November 9, 1978) there is a residential energy credit that can be claimed for heat pumps. It is in the form of an amendment to the Internal Revenue Code of 1954. In that code, Subpart A of Part IV, Subchapter A, Chapter 1 (relating to audits allowable) is amended by inserting after Section 44B the following new section:

> Sec. 44c Residential Energy Credit
>
> "(a) General Rule–In the case of an individual, there shall be allowed as a credit against the tax imposed by this chapter for the taxable year an amount equal to the sum of:
>
> (1) the qualified energy conservation expenditures, plus
>
> (2) the qualified renewable energy source expenditures."

Heat pumps qualify under (1) but not (2). For this case, the relevant part of the amendment continues as follows:

> "(b) Qualified expenditure–for purposes of subsection (a)–
>
> (1) Energy conservation–In the case of any dwelling unit the qualified energy conservation expenditures are 15 percent of so much of the energy conservation expenditures made by the tax payer during the taxable year with respect to such unit as does not exceed $2000."

Hence 15% of the heat pump's cost will be reported through the tax credit up to a maximum of $300. This is not enough to substantially augment a heat pump market where heat pumps are not already a sound economic investment.

As an example of the state legal influence, the complicated case of California illustrates the kind of cross-currents faced by a homeowner wanting to install a heat pump. If there is a solar supplement, it is eligible, until 1982, for a tax credit of 55% of the solar-related cost up to $3000 of credit ($5460 of cost). Unused credit may be carried over until 1982. This credit, however, is exclusive of the federal tax credit as both may not be applied to the same cost elements. Hence, the allocation made by the individual must account for his state and federal marginal tax brackets because state tax is deductible at the federal level, but not vice versa. If, on the other hand, supplemental electric resistance heaters are used (reasonable for the state's low winter heating requirements), recent California building regulations, in effect since February 1977, have a very restrictive effect. Electric resistance heating cannot exceed 10% of the building space heating requirement. Hence, the heat pump compressor would have to be oversized to minimize the auxiliary heating requirement when the temperature falls below the balance point.

MARKET PROSPECTS FOR THE HEAT PUMP

Most recently, prior to the Energy Tax Act and some of the most recent technical developments described in this book, Shaffer (1977) surveyed carefully the prospect for further growth in the heat pump market. The market for heat pumps appears to be divided between built-up units and unitary systems in proportions such as 1/6–5/6. Approximately 70% of the entire business is in small unitary pumps (5 tons and under), suitable primarily for residential use.

Tables 5-10 and 5-11 provide a breakdown of American Refrigeration Institute data on shipments of unitary systems during the recent 10-year period. The tables have been divided to show unit volume, dollar volume, kind of customer and also the unit and dollar volume in each of several size ranges. Further, both tables have been subdivided to show the demand for split systems (compressor/evaporator remote from condenser) in relation to the demand for packaged systems.

Very large unitary systems (100 tons and over) can be built, but more than 91% of the unitary heat pumps shipped in the seven years preceding 1974 were added to the total installed in residences. More than 95% of the unitary systems now manufactured (Table 5-11) are in the five-ton and under class and are suitable only for small structures.

As is shown by Tables 5-10 and 5-11, the business done in unitary heat pumps prior to 1973 was quite stable at about 100,000 units per year, at an annual volume of approximately $50 million. Widespread fuel shortages during the winter of 1972-73 followed by the oil embargo of 1973-74 appear to have produced a sudden upswing in the demand for heat pumps. The initial surge probably resulted from anxieties over gas and oil supplies triggered by the embargo; however, factors such as high fuel costs, nonavailability of new gas permits and/or other restrictions in many areas now seem to ensure a steady high growth rate. The actual figures for 1973–1975 show an average growth of 23%. There are reasons to discount, somewhat, the enormous surge in 1975–76; and, if long-term growth is projected at a more steady 20%/yr, we may expect annual

Table 5-10. Heat Pump Shipments
(thousands of units,[a] millions of dollars, single package, split systems, residential and commercial)
(American Refrigerator Institute Data)

	1967	1969	1970	1972	1973	1974	1975	1976	1980	1985
Single Package										
Thousands of units	49.0	59.9	62.5	64.0	73.0	77.1	71.8	112	146	240
Millions of dollars	23.5	30.1	–	33.6	44.0	57.0	57.0	91	–	–
Split Systems										
Thousands of units	30.7	37.4	35.2	33.6	47.0	61.4	94.8	163	304	810
Millions of dollars	14.1	17.4	16.3	17.2	20.0	24.0	35.0	60	–	–
Totals										
Thousands of units	79.7	97.3	97.7	97.6	120.0	130.5	166.6	275	450	1050
Millions of dollars	37.7	47.9	48.1	49.9	64.0	81.0	92.0	175	243	550
$/unit	473	492	492	511	533	504	1551	550	540	521
Estimated Units in Operation, Thousands	455	641	700	806	874	954	1056	1258	2443	5871
Single Package, thousands of units										
Residential	39.1	39.5	35.0	52.5	51.9	47.5	45.0	–	–	–
Commercial, large residential	8.6	19.4	26.4	8.2	18.3	23.9	23.1	–	–	–
Split Systems, thousands of units										
Residential	21.4	28.0	25.2	28.0	35.1	49.4	83.0	–	–	–
Commercial, large residential	8.3	8.8	8.4	4.2	8.8	7.2	9.0	–	–	–
Other	2.3	1.6	2.7	4.7	5.9	10.5	5.8	–	–	–

[a]The growth of units is based on current information combined with extrapolation of past history.

Table 5-11. Heat Pump Shipments (thousands of units, millions of dollars) (American Refrigeration Institute Data)

	1967		1969		1970	1972		1973	1974		1975	
	Units	$	Units	$	Units	Units	$	Units	Units	$	Units	$
Single Package												
2+ tons and under	16.4	5.48	18.8	6.59	20.0	20.1	7.11	19.8	18.5	7.0	16.5	5.0
2+ to 4– tons	22.3	10.13	30.2	14.50	32.5	26.7	13.12	36.4	38.2	20.8	37.8	21.1
4– to 5+ tons	9.15	6.30	9.4	6.91	8.9	16.0	11.40	15.2	18.1	13.9	14.9	13.0
5+ tons and over	1.15	1.57	1.6	2.09	1.1	1.2	1.90	1.6	2.3	4.3	2.5	5.3
Totals	49.0	23.48	60.0	30.09	62.5	64.0	33.53	73.0	77.1	46.0	71.7	44.4
Split Systems												
2+ tons and under	5.5	1.37	5.5	1.39	4.7	4.5	0.94	5.8	9.6	4.0	16.9	6.3
2+ to 4– tons	14.9	5.57	18.3	7.02	18.7	15.1	6.56	25.5	34.0	21.1	58.6	26.7
4– to 5+ tons	8.6	4.94	11.9	6.84	10.3	12.9	8.24	14.4	16.2	11.0	17.5	12.6
5+ tons and over	1.7	2.24	1.6	2.14	1.5	1.1	1.50	1.3	1.7	1.8	94.8	48.4
Totals	30.7	14.1	37.4	17.40	35.2	33.6	17.24	47.0	61.5	37.9		
Grand Totals[a]	79.7	37.7	97.3	47.9	97.7	97.6	49.9	120.0	138.5	83.9	166.5	92.8

[a]Slight discrepancy is due to error in rounding off.

shipments measured in millions of units by 1985. The dollar volume associated with this annual rate is hard to pin down because of inflationary factors, but recent trends suggest that the heat pump business (over $90 million in 1975 and approximately $150 million in 1976), may well exceed $500 million by 1985.

Over a long period, each heat pump unit shipped has been valued at $400-$600. The average for 1976 was $550. The number of large units included in the totals is so small that it hardly influences these figures, and we may conclude that the average small customer may have been offered equipment that would double the value of his electric heating dollar at about $1000 plus installation. When energy was cheap (< 2¢/kWh), this would not have been attractive unless the equipment was promoted on the basis of its value as an air conditioner. However, the rising trend in power costs, already averaging 4–5¢/kWh in the U.S., is expected to produce an increased interest in the use of heat pumps for all sorts of space-conditioning purposes. It is difficult to predict how strong an effect power costs will have because the value of the equipment is rising in accordance with the general inflationary trend. A further factor is the difficulty that the single individual has in obtaining capital funds as opposed to daily operating money. Real progress in the growth of the heat pump market would seem to depend on technological improvement. In the absence of any significant improvements in design, manufacture or installation techniques, we can expect the same lackluster growth that has characterized the past. There may be sudden spurts in response to crises of one kind or another, but it seems unlikely that equipment that merely doubles the value of a heating kWh could be widely marketed at a cost equal to two or three times the cost of installing another kilowatt of electric generating capacity.

Table 5-11 shows some of the trends in capacity categories of the market. Details in Tables 5-10 and 5-11 provide interesting insights into what is happening in the heat pump market. For example, Table 5-10 shows that in the residential area, the market share due to split systems grew from about one-third of the total in 1967 to slightly more than one-half in 1975. This may reflect to a great degree the purchase of heat pump systems to replace expensive electric furnaces and/or worn out or obsolete heating equipment of other types. However, even though some builders are offering new houses with two heat pumps to avoid the expense of ductwork and to give a two-zone effect, it seems likely that the use of split unitary systems will increase in single-family residential construction. The great popularity of remote compressor/condensers in the air-conditioning field—a market that is already 60 times the annual heat pump volume—is ample evidence of the general acceptability of a division of the system into an indoor segment and a remote outdoor compressor.

The heat pump market (as we see in Table 5-11) is overwhelmingly residential; 1 heat pump in 20 is not of small size (5 tons and under), and current and recent sales data indicate that residential heat pump installations account for two-thirds to four-fifths of the total annual value of heat pump shipments. Size distribution of the heat pump market estimated for 1979 by Gordian Associates (1978) is shown in Figure 5-6. The energy consumption figures show that residential heating accounts for 60% of the total heating energy requirement. If the use of heat pumps expands at all, unit sizes and sheer weight of numbers will produce a heat pump market that is primarily residence-oriented.

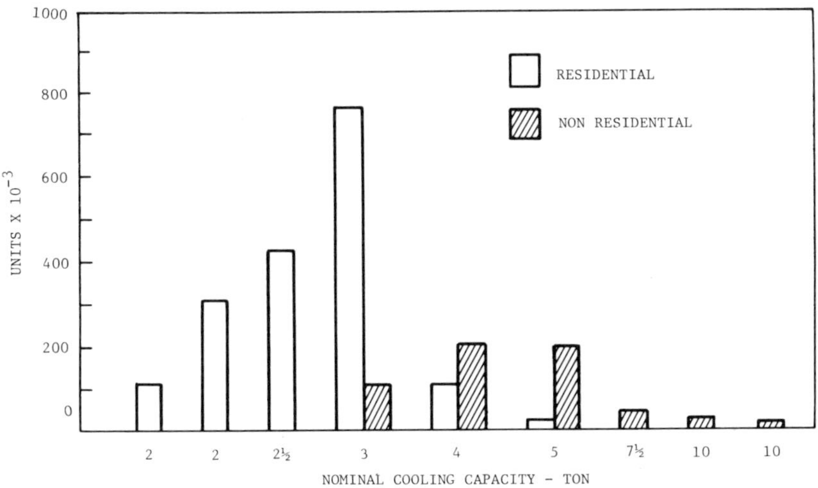

Figure 5-6. Size distribution of total potential market for advanced electric heat pumps, 1979 estimate (Gordian Associates, Inc., 1978).

To give some indication of the maximum possible demand for residential units, Table 5-12 proportions building data for recent years based on a total of 2,000,000 units and determines a potential market if all units are equipped with heat pumps on the assumption that a single residence would, on the average, require a 3.5-ton unit. Some assumptions were made regarding the sizes of larger units, but if wrong they will affect the total numbers rather than the total value, since the dollar figures were calculated from costs per ton of capacity, rather than from the total number of units shipped. The prices do not reflect recent upward trends and do not include any allowance for inflation. In any case, further large growth of the heat pump market will undoubtedly depend on realization

Table 5-12. Maximum Residential Heat Pump Market, based on 2,000,000 Residential Units Annually

	Housing Units	Potential Need for Heat Pumps: Heat Pumps			Average Size
Detached	959,400		959,400		3.5 tons
Town House	125,050		125,050		3.5 tons
Condominium	221,400		55,280		15 tons
Sales	1,305,850				
			1,084,450	@	3.5 tons
			55,280	@	15 tons
Town	102,500		102,500		3.5 tons
Garden Apartment	592,450		54,000		15 tons
			11,000		50 tons
Mid-Rise	10,250		260		100 tons
High-Rise	38,950		490		200 tons
Rentals	744,150				
			102,500	@	3.5 tons
			54,000	@	15 tons
			11,000	@	50 tons
			260	@	100 tons
Heat Pump Totals					
Unitary	1,186,950	3.5 tons		$1,780 million	
	109,280	15 tons		546 million	
				$2,326 million	
Built-up	674,000 net tons			$ 337 million	

of technological solution to the problems of their use in a cold climate. In these cases, the need is for seasonal COP of 3 rather than 2 which, as Shaffer (1977) pointed out, is not quite enough to pay for the investment and maintenance. A projection of heat pump market demands, assuming that little of this particular technological requirement will be met, has been cited by Gordian Associates (1978), as illustrated in Figure 5-7. The disparity over region does then persist in their projections, shown in more detail by Table 5-13.

Such estimates, based on a continuation of current trends within a substantially unchanged technological framework, are not reliable if substantial changes do occur. For these particular estimates, the area subject to error clearly lies in the populous U.S. Northeast, if there is a spread of heat pump economic viability to colder climates. The use of cheap,

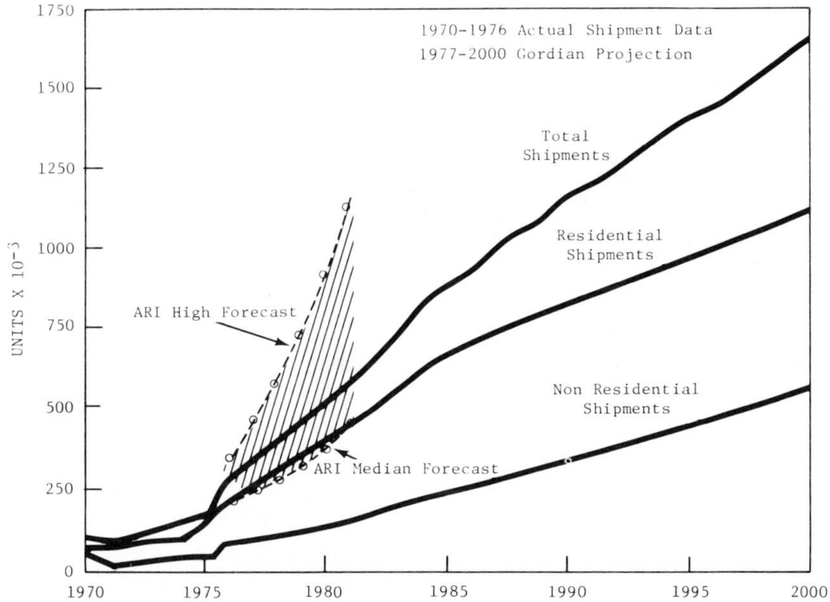

Figure 5-7. Historical and projected domestic market for standard electric heat pumps (Air-Conditioning and Refrigeration Institute; Gordian Associates, Inc., 1978).

Table 5-13. Potential (residential building) Market for Electric Heat Pump, 1981 (Gordian Associates, Inc., 1978)

	Northeast	North Central	South	West	Totals
Projected Total New Construction (units x 10^{-3})	228	367	788	367	1750
Central Air Conditioning Saturation (%)	22.7	48.3	86.0	46.6	–
Total New Construction Potential Market (units x 10^{-3})	52	178	678	171	1079
Replacement Market[a] (units x 10^{-3})	16	97	398	100	611
Conversion Market[b] (units x 10^{-3})	4	7	13	7	31
Total Potential Market (units x 10^{-3})	72	282	1089	278	1721

[a]Replacement market based on warm air installations.

[b]Conversion market based on 1975 actual heat pump conversions.

low-capacity thermal storage to shift the diurnal duty cycle, relieving the need for heat pump operation against severely cold nighttime extremes, would go far toward raising seasonal COP. Another way would be to relieve the very high heating capacity requirement through building orientation, exterior design and massive thermal insulation. For example, there has been some successful experimentation by a small independent builder aimed at reducing heating/cooling requirements by 80%. In Alexandria, Virginia, home building recommendations, originally developed in 1945, were applied to achieve this result with an increase in building costs that was less than $2000 (*Washington Post*, 5 March 1979). When space thermal conditioning requirements are reduced by a factor of 5, capital investment required for all components of the heat/cooling system can drop by almost the same amount. For example, thermal storage requirements can be reduced by a combination of capacity and upper operating temperature to meet reduced maximum thermal requirements. The large drop in fuel and electricity requirements would allow economy for heat pumps in cold areas simply for their convenience.

Due to either, or some combination of the two possible kinds of basic influences posed above, we could expect the cold U.S. areas to come up to the South's level of per capita demand for heat pumps, in the unit terms of Table 5-13. However, in time, total operational capacity of the heat pump market might actually decline under the influence of such strong energy-conservative measures in home building. Such potential cross-currents indicate a level of reserve that should be expressed regarding the simple trend estimates.

REFERENCES

Air Cond. Heat. Refrig. News (March 17, 1975).

Ambrose, E. R. "The Heat Pump: Performance Factor, Possible Improvements," *Heat. Pip. Air Cond.* 77 (May 1974).

Auxer, W. L. "Development of a Stirling Engine Powered Heat Activated Heat Pump," *12th Intersociety Energy Conversion Engineering Conference Washington, D.C.,* 28 Aug-2 Sept (1977), p. 397.

Beale, W., J. Raush and S. Lewis. "Design, Details and Performance Characteristics of Some Free Piston Stirling Engineer," *8th Intersociety Energy Conversion Engineering Conference, University of Pennsylvania, Philadelphia, PA* (1973), p. 190.

Blake, P. J., and W. C. Gernert. "Load and Use Characteristics of Electric Heat Pumps in Single-Family Residences," Westinghouse Electric Corp., EA-793, RP 432-1 Pittsburgh, PA (1978).

Boxer, H. "Rotary Powered Compressors," *Second Annual Heat Pump Technology Conference, Oklahoma State University, Stillwater, OK* (1976).

Brown, G. V. "Magnetic Heat Pumping Near Room Temperature," *J. Appl. Physics* 47(8):3673 (1976).

Brown, G. V. "Magnetic Stirling Cycles–A New Application for Magnetic Materials," *IEEE Trans. Magnetics* 13(5):1146 (1977).

Brown, G. V. "Magnetic Heat Pumping," U.S. Patent 4,069,028 (January 17, 1978).

Cawley, R. E., and D. M. Pfarrer. "Part Load Efficiency Advantages of Two-Speed Refrigerant Compressors," *Proc. Conf. on Improving Efficiency and Performance of HVAC Equipment and Components for Residential and Small Commercial Buildings, Purdue University* (1974), p. 42.

Cheney, L. W., "A New Approach to Compressor Modulation," *Proc. 1974 Purdue Compressor Technical Conference*, 10-12 July 1974, W. Lafayette, IN.

Christian, J. E. "Unitary Air-to-Air Heat Pumps," Argonne National Laboratory ANL/CES/TE 77-10 (1977).

Colosimo, D. D. "Gas and Electric Heat Pump Alternatives," Mechanical Technology Incorporated, Latham, NY (1978).

Curran, H. M. "Assessment of Solar-Powered Cooling of Buildings," ERDA Report No. NSF-RA-N-75-012 (1975).

Davis, E. S., and L. C. Wen. "Solar Heating and Cooling System for Buildings: Technology and Selected Case Studies," Jet Propulsion Laboratory, California Institute of Technology, Pasadena, CA (1975).

Domingorena, A. A. "Performance Evaluation of a Low-First-Cost, Three-Ton, Air-to-Air Heat Pump in the Heating Mode," Oak Ridge National Laboratory ORNL/CON-18, Oak Ridge, TN (1978).

Drucker, E. E., and M. Ucar. "Computer Model for a Solar-Assisted Closed Liquid Loop Heat Pump System," paper presented at ASHRAE meeting, Dallas, TX, February 1-6, 1976.

Drucker, E. E., J. LaGraff, H. Card, W. Fleming, M. Ucar and F. Lansing. "Commercial Building Unitary Heat Pump System with Solar Heating," NSF Report on Grant GI-43895 (1976).

Fallon. J., and T. Iles. "High Temperature Brayton Cycle Heat Pumps," paper presented at the Franco-American Symposium on Energy Conservation in Industry, Paris, October 18, 1978.

French, R. L., E. S. Davis and R. K. Baerwald. "Design and Economics of Solar Augmented Heat Pumps and Thermal Storage for Southern California Homes," Jet Propulsion Laboratory, California Institute of Technology, Pasadena, CA (1976).

Gates, R. R., C. F. Sepsy, and G. D. Huffman. "Heat Transfer and Pressure Loss in Extended Surface Heat Exchangers Operating Under Frosting Conditions, Part II, Loss Analysis and Corrections," *ASHRAE Trans.* 73 (Part II) 1.2.1 (1967).

Gordian Associates, Inc. "Heat Pump Technology: A Survey of Technical Developments, Market Prospects and Research Needs," Gordian Associates, Inc., New York, NY, for Department of Energy EX-76-C-01-2121 (1978).

Gorzelnik, E. F. "Heat Pump Water Heater Makes Debut," *Elec. World* 191(8):80 (1979).

Holman, A. S., and V. R. Brantley. "ACES Demonstration, Startup and Performance Report," Oak Ridge National Laboratory ORNL/CON-26 (1978).

Hudlicky, M. "Chemistry of Organic Fluorine Compounds" (Elmsford, NY: Pergamon Press, Inc., 1961).

Huffman, G. P., and Sepsy, C. F. "Heat Transfer and Pressure Loss in Extended Surface Heat Exchangers Operating Under Existing Conditions, Part II: Data Analysis and Correction," *ASHRAE Trans.*, 73, Part II, P. I.3.1 (1967).

Johnson, J. H. "Hermetic Motor Efficiency," paper presented at the Conference on Improving the Efficiency in HVAC Equipment and Components, Purdue University, October 7-8, 1974.

Kays, W. M., and A. L. Landon, "Compact Heat Exchangers," (New York: McGraw-Hill Book Co., 1964).

Kelly, G. E., and J. Bean. "Dynamic Performance of a Residential Heat

Pump," *Proc. 2nd Annual Heat Pump Technol. Conf., Oklahoma State University, Stillwater, OK* (1976).

Kelly, G. E., and W. H. Parken. "Method of Testing, Rating and Estimating the Seasonal Performance of Central Air Conditioners and Heat Pumps Operating in the Cooling Mode," NTIS PB 280 107 (1978).

Kirschbaum, H. S., and S. E. Veyo. "Investigation of Methods to Improve Heat Pump Performance and Reliability in a Northern Climate," Vols. I and III, EPRI Report EM-319 (1977).

Lawrence, L. R. "Heating and Cooling Technologies for Natural Gas Conservation," *Proc. 13th Intersoc. Energy Conversion Eng. Conf.,* San Diego, CA (1978), p. 811.

Liebowitz, H., and D. Colosimo. *Heat Pumping Technologies,* Vol. II, *Technical and Economic Screening* (Latham, NY: Mechanical Technology, Inc., 1978).

"Lennox 2-Speed Compressor is Major Industry Breakthrough," *Lennox News* 27(4) (1973).

Lovvorn, N. C. "Heat Pump Compressor Reliability," *Air Cond. Heat. Refrig. News* (January 27, 1975).

Midgley, T., Jr., and A. L. Henne. *Ind. Eng. Chem.* 24:641 (1932).

"Exxon to Develop a Fuel Saver for Electric Motors," *New York Times* 1 (May 19, 1979).

"Fedders Agrees to Recall and Fix 40,000 Pumps," *New York Times* A20 (February 23, 1979).

O'Neill, D., J. Carney and E. Hirst. "Regional Analysis of Residential Water Heating Options: Energy Use and Economics," Oak Ridge National Laboratory ORNL/CON-31 (1978).

Orange, D. "Science, Product Quality Keys to Continued Heat Pump Success," *Air Cond. Refrig. Bus.* 112 (January 1977).

Parish, H. C., and C. F. Sepsy. "A Numerical Analysis of Frost Formation Under Forced Convection," *ASHRAE Trans.* 78 (Pt. 1) 236 (1972).

Parker, W. H., R. N. Beausoleil and G. E. Kelly. "Factors Affecting the Performance of a Residential Air-to-Air Heat Pump," paper presented at at the ASHRAE meeting, Chicago, IL, February, 1977.

Patton, P. "What Dr. Runsdorf Knows (and the Government Doesn't) about Legionnaire's Disease," *New York Mag.* 30 (January 29, 1979).

Pickelsimer, E. A. "Heat Pumps–Substitute for Out-Moded Fossil-Fueled Systems," Lockheed Missiles and Space Company, Huntsville, AL (1977).

Pickelsimer, E. A. Personal communication (1979).

Richards, W. D. C., and W. L. Auxer. "Performance of a Stirling Engine Powered Heat Activated Heat Pump," *Proc. 13th Intersoc. Energy Conversion Eng. Conf.*, San Diego, CA (1978), p. 823.

Shaffer, L. H. "The Heat Pump Commercial Forecasts," Business Communications Company (1977).

Stewart, R., J. Healey, B. Murphy and J. Scott. "Solar Energy and the Heat Pump in a Northern Climate," EPRI EA-407 (1977).

Swenson, P. F., and R. K. Rose. "Development of a High Seasonal Performance Factor Gas Heat Pump," *Proc. 12th Intersoc. Energy Conversion Eng. Conf.,* Washington, DC (1977), p. 390.

Visher, S. S. "Climatic Atlas of the United States," *Harvard University Press*, 1954, 403 pp.

"Exxon Corporation Seeking to Acquire Reliance Electric Company for Aid in Development of a New Energy Technology," *Wall Street J.* 3 (May 21, 1979).

"Fedders Consents to FTC Order, Will Pay for Repairs to About 40,000 Heat Pumps," *Wall Street J.* 10 (February 23, 1979).

"Fuel Costs Cut Sharply in Six New Virginia Homes," *Washington Post* C-1 (March 5, 1979).

West, J. E. "Defrosting of Air-to-Air Heat Pumps," *Proc. 2nd Annual Heat Pump Technol. Conf.*, Oklahoma State University, Stillwater, OK (1976).

Williams, T. E. "Rankine Powered Industrial Heat Pump," paper presented at the Franco-American Symposium on Energy Conservation in Industry, Paris, October 18, 1978.

INDEX